The Future of U.S. Nuclear Weapons Policy

Committee on International Security and Arms Control

National Academy of Sciences

NATIONAL ACADEMY PRESS
Washington, D.C. 1997

NATIONAL ACADEMY PRESS • 2101 Constitution Avenue, N.W. • Washington, DC 20418

NOTICE: The project that is the subject of this report was approved by the Council of the National Academy of Sciences. The Committee on International Security and Arms Control is a standing committee of the National Academy of Sciences, whose members were chosen for their special competences and with regard for appropriate balance.

This report has been reviewed by a group other than the authors according to procedures approved by a Report Review Committee consisting of members of the National Academy of Sciences, the National Academy of Engineering, and the Institute of Medicine.

The National Academy of Sciences is a private, nonprofit, self-perpetuating society of distinguished scholars engaged in scientific and engineering research, dedicated to the furtherance of science and technology and to their use for the general welfare. Upon the authority of the charter granted to it by the Congress in 1863, the Academy has a mandate that requires it to advise the federal government on scientific and technical matters. Dr. Bruce M. Alberts is president of the National Academy of Sciences.

The National Academy of Engineering was established in 1964, under the charter of the National Academy of Sciences, as a parallel organization of outstanding engineers. It is autonomous in its administration and in the selection of its members, sharing with the National Academy of Sciences the responsibility for advising the federal government. The National Academy of Engineering also sponsors engineering programs aimed at meeting national needs, encourages education and research, and recognizes the superior achievement of engineers. Dr. William A. Wulf is president of the National Academy of Engineering.

The Institute of Medicine was established in 1970 by the National Academy of Sciences to secure the services of eminent members of appropriate professions in the examination of policy matters pertaining to the health of the public. The Institute acts under the responsibility given to the National Academy of Sciences by its congressional charter to be an adviser to the federal government and, upon its own initiative, to identify issues of medical care, research, and education. Dr. Kenneth I. Shine is president of the Institute of Medicine.

The National Research Council was organized by the National Academy of Sciences in 1916 to associate the broad community of science and technology with the Academy's purposes of furthering knowledge and advising the federal government. Functioning in accordance with general policies determined by the Academy, the Council has become the principal operating agency of both the National Academy of Sciences and the National Academy of Engineering in providing services to the government, the public, and the scientific and engineering communities. The Council is administered jointly by both Academies and the Institute of Medicine. Dr. Bruce M. Alberts and Dr. William A. Wulf are chairman and vice-chairman, respectively, of the National Research Council.

This project was made possible with funding support from the John D. and Catherine T. MacArthur Foundation. Any opinions, findings, conclusions, or recommendations expressed in this publication are those of the author(s) and do not necessarily reflect the view of the organizations or agencies that provided support for the project.

Available in limited quantities from:

The Committee on International
 Security and Arms Control
National Academy of Sciences
2101 Constitution Avenue, N.W.
Washington, DC 20418
cisac@nas.edu

Additional copies are available for sale from:

National Academy Press
2101 Constitution Avenue, N.W.
Box 285
Washington, DC 20055
1-800-624-6242
(202) 334-3313 (in the Washington metropolitan area)
http://www.nap.edu

Library of Congress Catalog Card Number 97-68120
International Standard Book Number 0-309-06367-1

Foreword

Unlike most National Research Council committees, which are formed to carry out a particular study and then dissolved when their task is complete, the Committee on International Security and Arms Control (CISAC) is a standing committee of the National Academy of Sciences. CISAC was created in 1980 to bring the Academy's scientific and technical talent to bear on crucial problems of peace and security. The committee's objectives are to engage scientists in other countries in dialogues that build a common understanding of security issues and work toward common solutions to arms control and security problems, to develop recommendations and other initiatives on scientific and technical issues affecting international security and cooperation, to respond to requests from the U.S. government for analysis and advice on these issues, and to inform and foster the interest of scientists and engineers in international security problems.

The committee's rotating membership includes scientists, engineers, and policy analysts. John P. Holdren (Harvard University) serves as chair of the committee, with John Steinbruner (The Brookings Institution) as vice-chair.

Together, CISAC's members have many decades of experience in nuclear policy, many in senior government positions, dating back to the Manhattan Project (see Appendix A for biographies). All of them are currently involved in security affairs on at least a part-time basis. This report reflects the collective technical and political judgment of these individuals. Although grounded in technical assessments wherever possible, the committee acknowledges that there are points where the analysis results from its discussions and joint study of the issues rather than from "facts" alone. As my predecessor, Frank Press, said of CISAC's 1991 study: "Rather than developing new ideas, the study's greatest value lies in the remarkable degree of consensus that the group was able to achieve on a wide array of important security issues" (*The Future of the U.S.-Soviet Nuclear Relationship*, p. vii). Some of CISAC's members might have preferred more or less ambitious recommendations on some issues, but in the end the committee agreed on a comprehensive program that would transform the roles that nuclear weapons play in the national security policy of the United States.

Major General William F. Burns (USA, ret.) chaired this study for CISAC. He has been engaged in many aspects of nuclear policy over the years; one of his first assignments was to an artillery battalion armed with tactical nuclear weapons on the front lines of NATO and, after a distinguished military career, one of his last government assignments was as director of the Arms Control and Disarmament Agency. CISAC is deeply indebted to him for accepting this demanding task and seeing it to completion with patience, good humor, and unflagging intellectual engagement in shaping the committee's conclusions and recommendations. Every member of CISAC contributed to the text of the study; Steve Fetter, John P. Holdren, Spurgeon Keeny, and Wolfgang K. H. Panofsky undertook particularly heavy drafting assignments.

The committee also is grateful for the assistance it received in the course of the study. CISAC's director, Jo Husbands, was indispensable as usual in her contributions to the organization, coordination, drafting, and editing of the report. Her professionalism, tact, and willingness to extend herself on behalf of members of CISAC reflect great credit on her as a member of the Academy's senior staff. Michael Mazarr served as a consultant in the early stages of the study and contributed significantly to its formulation and development. La'Faye Lewis-Oliver provided invaluable administrative support and budget-stretching skills.

The report has the unanimous endorsement of all CISAC members, with the exception of Joshua Lederberg who was engaged in another major CISAC project on biological weapons issues and was unable to participate in the study process.

BRUCE ALBERTS
President
National Academy of Sciences

Contents

EXECUTIVE SUMMARY 1
  The Current Situation and the Reasons for Further Change, 1
  Nuclear Deterrence Past and Future, 3
  A Two-Part Program of Change, 4
  Building a Regime of Progressive Constraints, 6
  Prohibition of Nuclear Weapons?, 8

1 WHY CHANGE U.S. NUCLEAR WEAPONS POLICY? 11
  The Problem and the Prospects in Summary, 12
  Nuclear Weapon Dilemmas and Dynamics, 23
  The Case for Post-Cold War Reductions and Transformations, 26
  Orientation and a Caution, 30

2 CURRENT U.S. NUCLEAR WEAPONS POLICY 33
  The U.S.-Russian Interaction, 33
  The Other Nuclear Weapons States: China, France,
    and the United Kingdom, 46
  Nuclear Weapons Policy and Nonproliferation, 47
  Conclusion, 56

3 A REGIME OF PROGRESSIVE CONSTRAINTS 58
  An Immediate Step: To 2,000 Deployed Strategic Warheads, 59
  Further Transformation of the U.S.-Russian Interaction, 60
  Nuclear Reductions and Nonproliferation, 66
  Nuclear Force Reductions: How Low Can We Go?, 75

4 PROHIBITION OF NUCLEAR WEAPONS 85
    The Benefits and Risks of Nuclear Disarmament, 86
    Prerequisites for Nuclear Disarmament, 88
    Routes to Nuclear Disarmament, 92
    Conclusions, 97

APPENDIXES

A   Biographical Sketches of Committee Members 101
B   The Buildup and Builddown of Nuclear Forces 105

Executive Summary

The debate about appropriate purposes and policies for U.S. nuclear weapons has been under way since the beginning of the nuclear age. With the end of the Cold War, the debate entered a new phase, propelled by the post-Cold War transformations of the international political landscape and the altered foreign policy challenges and opportunities that these changes are bringing about. This report—based on an exhaustive reexamination of the issues addressed in the committee's 1991 report on *The Future of the U.S.-Soviet Nuclear Relationship*—describes the state to which U.S. and Russian nuclear forces and policies have evolved since the Cold War ended, the reasons why further evolution is desirable, and the shape of a regime of progressive constraints responsive to these reasons. It concludes with a discussion of the conditions and means under which, in the longer term, it could become desirable and feasible to prohibit the possession of nuclear weapons altogether.

THE CURRENT SITUATION AND THE
REASONS FOR FURTHER CHANGE

The first Strategic Arms Reduction Treaty (START I), signed in 1991 as the Cold War was ending and now being implemented by both the United States and Russia, will reduce the number of strategic nuclear warheads deployed by the two countries from 13,000 and 11,000, respectively, to about 8,000 each. START II, signed in 1993 and ratified by the United States in early 1996 but not yet (as of this writing) ratified by Russia, would further limit the number of deployed strategic warheads to 3,000 to 3,500 on each side. At the Helsinki summit in March 1997 Presidents Clinton and Yeltsin agreed to seek a START III treaty with a level of 2,000 to 2,500 deployed strategic nuclear warheads. Unilateral initiatives since the Cold War began winding down have also reduced very substantially the numbers of deployed nonstrategic warheads, especially on the U.S. side. In addition, nuclear testing has ended, the United States and Russia have agreed not to target their missiles against each other on a day-to-day basis, and production of weapons-grade fissile material has stopped in the United States and is expected to stop soon in Russia.

These actions have unambiguously halted and reversed the bilateral nuclear competition that was the most conspicuous characteristic of the Cold War's military confrontation but, unfortunately, have not sufficiently altered the physical threat that these weapons pose. The reduced forces could still inflict catastrophic damage on the societies they target or could target, and the thousands of nondeployed and nonstrategic nuclear warheads not addressed by the START process and likely to be retained without further agreements will pose substantial risks of breakout, theft, or unauthorized use. In addition, the United States and its NATO allies retain their Cold War "weapons of last resort" doctrine that allows the first use of nuclear weapons if deemed necessary to cope with nonnuclear attacks, and Russia has announced that it is abandoning the Soviet Union's no-first-use pledge in order to adopt a position similar to NATO's.

The basic structure of plans for using nuclear weapons appears largely unchanged from the situation during the Cold War, with both sides apparently continuing to emphasize early and large counterforce strikes and both remaining capable, despite reductions in numbers and alert levels, of rapidly bringing their nuclear forces to full readiness for use. As a result, the dangers of initiation of nuclear war by error (e.g., based on false warning of attack) or by accident (e.g., by a technical failure) remain unacceptably high. (On the Russian side, the dangers of erroneous, accidental, or unauthorized nuclear weapons use may be even higher than during the Cold War because of subsequent deterioration of the military and internal-security infrastructure and of morale.) The continuing competitive assumptions underlying some official discussions of the U.S.-Russian nuclear relationship, when coupled with the postures of the forces and the potential for destabilizing deployments of ballistic missile defenses, pose the risk that the arms control fabric woven during the Cold War and immediately thereafter could unravel.

In addition, continued actions by the United States and Russia to reduce their nuclear arsenals—and to reduce the roles assigned to those arsenals—are needed to help bring the other declared and undeclared nuclear weapons states into the arms reduction process and to strengthen the global nonproliferation regime. The effectiveness of that regime depends on the full support and cooperation of a large number of nonnuclear weapons states in the maintenance of a vigorous International Atomic Energy Agency with the inspection powers and resources to do its job, in the implementation of effective controls on the transfer of sensitive technologies, and in the creation of transparency conditions conducive to building confidence that proliferation is not taking place. The degree of commitment of the nonnuclear weapons states to these crucial collective efforts will surely depend at least in part on impressions about whether the nuclear weapons states are working seriously on the arms reduction part of the global nonproliferation bargain.

NUCLEAR DETERRENCE PAST AND FUTURE

During the Cold War, nuclear deterrence was the bedrock of U.S strategy for preventing both nuclear war and major conventional war because a more effective alternative was not apparent: the adversarial U.S.-Soviet relationship made it seem imprudent to rely on good intentions to preclude nuclear attack or massive conventional assault; the character of nuclear weapons and the diverse means for delivering them meant that attempts to defend the United States or its allies against nuclear attacks on their populations could be overcome with much less effort than would have to be invested in the defenses; highly survivable basing modes for significant parts of each side's nuclear forces made it impractical to execute a disarming first strike even if conflict seemed imminent; and concern about the powerful conventional forces of the Warsaw Treaty Organization (and, in Asia, those of China and North Korea) motivated the United States and its allies to adopt a first-use-if-necessary nuclear posture to deter large-scale conventional attacks.

But nuclear deterrence itself was (and is) burdened with an array of dilemmas and dangers. For example, deterrence is likely to succeed only if there are credible plans for what to do if it fails, but constructing such plans is exceedingly difficult, and attempts to make the threat of nuclear retaliation credible can be seen as aggressive advantage seeking by the other side. This raises tensions, stimulates arms races, or increases the chance of nuclear war from crisis instability or accident. In addition, the assertion by some countries of a need and a right to have a nuclear deterrent may encourage additional countries to assert the same need and right, leading to further nuclear proliferation.

This committee has concluded that the dilemmas and dangers of nuclear deterrence as practiced by the United States in the past can and should be alleviated in the post-Cold War security environment by confining such deterrence to the *core function* of deterring nuclear attack, or coercion by threat of nuclear attack, against the United States or its allies. That is, the United States would no longer threaten to respond with nuclear weapons against conventional, chemical, or biological attacks. Given adequate conventional forces, the active and conspicuous role given to nuclear weapons during the Cold War can be greatly reduced without significant adverse effect on the probability of major war or on this country's ability to deal effectively with regional conflicts where its vital interests and those of its allies are at stake. The committee believes that Russia and the other nuclear weapons states can be persuaded to reach a comparable conclusion.

In all likelihood the United States will consider it necessary to continue to rely on the core function of nuclear deterrence as long as nuclear weapons continue to exist in the possession of states that might consider using them against this country or its allies. The committee assumes that some—although it is hoped not all—other nuclear weapons states will similarly consider it necessary to retain some nuclear weapons for "core deterrence." But the size and scope of the efforts deemed necessary by the United States and others to fulfill the core function presumably will shrink in parallel with what the committee hopes is the declining plausibility that any state would consider mounting a nuclear attack on anyone. Moreover, there are strong reasons to make every effort to hasten the arrival of international conditions in which threats of nuclear attack are simply no longer thinkable, so that the practice of deterrence with all its dilemmas and dangers would no longer be necessary.

As long as nuclear weapons exist, this very existence will exert a deterrent effect—*existential deterrence*—against unrestricted conventional war among the major powers, since it will be recognized that, in a world with nuclear weapons, such conflicts might well lead to their use, with intolerable destruction as the result. Indeed, even the existence of the idea of nuclear weapons—more specifically, the ability of many states to make them—is enough to create an existential deterrent effect against large-scale conflicts of all kinds. That is not to say that this effect would necessarily always be sufficient to prevent conflict in the future, any more than it has always been in the past. But it could provide a *part* of the assurance required, in an international system much different than today's, that all-out wars are unlikely to occur.

A TWO-PART PROGRAM OF CHANGE

If only the core function of nuclear weapons retains validity, fundamental changes in the nuclear force structures and operational practices of the major nuclear powers become both possible and desirable. Accordingly, the committee has concluded that the United States should pursue a two-part program of change in its nuclear weapons policies.

- The first part of the program is a near- and midterm set of force reductions—together with accompanying changes in nuclear operations and declaratory policies and with measures to increase the security of nuclear weapons and fissile materials worldwide—to diminish further confrontational and potentially destabilizing aspects of force postures, to reduce the risks of erroneous, unauthorized, or accidental nuclear-weapons use, and to help curb the threat of further nuclear proliferation. In their early phases these measures are largely bilateral ones between the United States and Russia, and close cooperation between the two countries is essential for success.

- The second part of the program is a long-term effort to foster international conditions in which the possession of nuclear weapons would no longer be seen as necessary or legitimate for the preservation of national and global security.

Nuclear force reductions and changes in nuclear operations would increase U.S. and global security in important ways.

- First, reducing U.S. and Russian nuclear forces and revising operations for the mission of fulfilling only the core function will decrease the continuing risk of accidental, erroneous, or unauthorized use of nuclear weapons for several reasons. Smaller arsenals will be easier to safeguard and protect from accident, theft, and unauthorized use, not only by virtue of reduced numbers of weapons to monitor at a smaller number of sites but also by permitting retention of only those weapons with the most modern safety and security features. Reducing alert rates, decreasing capacities to use nuclear weapons quickly and with little warning, abandoning plans for the rapid use of nuclear weapons, and deploying cooperative measures to assure states that forces are not being readied for attack should reduce the probability and consequences of erroneous nuclear weapons use—for example, on false warning of attack. (Of course it is extremely important to take care that reductions in deployed nuclear warheads—and dismantlement of the warheads made surplus as a result—do not lead to countervailing increases in the dangers of theft and unauthorized use as a consequence of inattention to the challenges of safe storage of these weapons and the nuclear materials removed from them.)
- Second, further reductions will bolster the nuclear nonproliferation regime. U.S.-Russian nuclear arms reductions will not in themselves dissuade a state bent on acquiring nuclear weapons; today's undeclared nuclear powers and would-be proliferators are driven above all by regional security concerns. In such cases, the denial of material and technical resources and a combination of political and economic incentives and disincentives provide the greatest leverage. But U.S. and Russian progress in arms reductions helps shore up global support for antiproliferation measures; and lack of such progress can strengthen the influence of those arguing for nuclear weapons acquisition in countries where this is under internal debate.
- Third, continued actions by the United States and Russia to reduce their nuclear arsenals—and the roles and missions assigned to those arsenals—will help persuade the other declared and undeclared nuclear weapons states to join the arms control process. At planned START II levels, for example, under which it is estimated that the United States and Russia each would retain a total of about 10,000 nuclear warheads, deployed and in reserve, the other nuclear powers have little motivation to submit their much smaller arsenals to any form of control.

BUILDING A REGIME OF PROGRESSIVE CONSTRAINTS

The program that the committee recommends would shift the focus of U.S. nuclear policy. While preserving the core function of deterring nuclear aggression, nuclear forces would be reduced, their roles would be more narrowly defined, and increased emphasis would be placed on achieving higher standards of operational safety.

Building on past nuclear arms control agreements and the anticipated START III agreement, future bilateral U.S.-Russian negotiations should center on specific means to achieve these goals. The first step is to encourage the Russian Duma's ratification of START II by beginning now to discuss a START III agreement limiting the number of deployed strategic warheads to about 2,000 on each side.

Establishing progressive constraints on nuclear operations is equally urgent; additional efforts should be pursued in parallel with, but not linked to, discussions of a START III agreement. Such constraints would include programs to reduce alert levels further and progressively to reorient nuclear doctrine away from the requirement to plan for rapid, massive response. Limits on ballistic missile defenses consistent with the Anti-Ballistic Missile (ABM) Treaty would be maintained.

A continuing high priority effort is also needed to improve the protection of nuclear weapons and fissile materials in Russia. Joint U.S.-Russian work along these lines, which has been going on since 1991 under the Nunn-Lugar Cooperative Threat Reduction Program, complements and strengthens arms reductions and other changes in nuclear policies. (Because this committee and other NRC committees have recently offered detailed analysis and recommendations on this subject in other reports, the committee does not treat it in detail here.)

During the Cold War, reducing the risk of a surprise attack appeared to be more important than the risks generated by maintaining nuclear forces in a continuous state of alert. With the end of that era, the opposite view is now more credible. This has important implications for U.S. nuclear policy and calls for dramatically reduced alert levels. Elimination of continuous-alert practices should be pursued as a principal goal in parallel with, but not linked to, START III. As a related confidence-building measure, the United States and Russia should adopt cooperative practices to assure each other that they are not preparing for a nuclear attack.

With the Cold War over, planning to retaliate massively against a nuclear attack is not the appropriate basis for making responsible decisions regarding the actual use of nuclear weapons. Operational doctrine regarding the magnitude and timing of any actual retaliation in response to a nuclear attack should be revised. The United States should adopt a strategy that would permit much more selective targeting options and that would be based neither on predetermined prompt attacks on counterforce targets nor on automatic destruction of cities. The presumption instead would be that nuclear weapons, if they were ever to be used, would be employed against targets that would be designated in response to immediate circumstances—in the smallest possible numbers. Some changes in this direction have begun, but the move to a more flexible planning system should be accelerated.

Together, positive and negative security assurances and guarantees have been a useful policy tool to ensure that friends and allies of the United States are not penalized by foregoing nuclear weapons. The United States could do more, however, to make negative security assurances and guarantees serve nonproliferation interests. Most important, the United States should adopt no-first-use of nuclear weapons as its declaratory policy at an early date. Changing to a no-first-use policy will, of course, require consultation with allies to reassure them that the United States will meet, by nonnuclear means, its obligations to come to their aid in the event of a nonnuclear attack on them.

Efforts to ban nuclear weapons from specific regions or environments strengthened nonproliferation in the past and helped to limit the perceived utility of nuclear weapons. The United States should continue to support these agreements and sign them without reservations that undermine their basic purpose, consistent with the unequivocal no-first-use policy recommended above. A new nuclear weapon free zone in Central Europe would, for example, offer immediate security advantages to Russia as well as NATO.

The committee has concluded that the changed international security environment makes possible further reductions in nuclear armaments. After the reductions envisioned in a START III accord, reduction to about 1,000 *total* warheads each for the United States and Russia would be a logical next step. (All nuclear warheads—regardless of type, function, stage of assembly, associated delivery system, or basing mode—would then be included in the negotiated limits.) A force of this size could effectively maintain the core function against the most challenging potential U.S. adversaries under any credible circumstances. This reduction process must ensure stability at each rung of the ladder, requiring survivable nuclear forces not at risk from a first strike.

Verifying limits on total nuclear warheads is substantially more difficult than verifying limits on their delivery vehicles. Verifying numbers of nondeployed and nonstrategic warheads, in particular, would require transparency measures regarding the production, storage, and dismantling of nuclear warheads, as well as a mechanism for exchanging and verifying information about the location and status of warheads. Since nuclear weapons can be small and portable and not easily detectable by technical means, however, a regime that would provide high confidence of locating a small number of hidden warheads would be extremely difficult to achieve. Even an imperfect verification regime would greatly reduce the uncertainties in present U.S. estimates of the number of Russian warheads.

Fulfilling the goals of current arms control initiatives and successfully providing for much deeper reductions will also require improved standards of accounting, transparency, and physical security for fissile materials. Efforts to control fissile materials must address not only the problems presented by military stockpiles but also by civilian use of such materials, in particular plutonium produced by reprocessing. A fissile material cutoff would be a significant nonproliferation measure and should continue to be strongly supported by the United States.

The ABM treaty will continue to play a crucial role in a world in which the numbers of nuclear weapons are drastically reduced and the role of nuclear weapons is restricted to the core function. Maintaining and enhancing its integrity in light of changes in offensive nuclear capabilities will require periodic evaluation. The focus of the U.S. ballistic missile defense research and development program should be to field a mobile system capable of defending relatively small areas against projected theater ballistic missile threats, which the committee believes will remain limited to a range of roughly 1,000 kilometers for some time.

The achievement of U.S.-Russian reductions to a mutually agreed level of about 1,000 total warheads each should not represent the final level for nuclear arms reductions. There will still be powerful reasons to continue down to a level of a few hundred nuclear warheads on each side, with the other three declared nuclear powers at lower levels, or with no remaining nuclear forces.

The small numbers of nuclear weapons presumably now held by the undeclared nuclear states—India, Israel, and Pakistan—would become a key issue when the United States and Russia, as well as the other declared nuclear powers, consider reductions to very small numbers of warheads. High priority should be given to diplomatic strategies tailored to the security perceptions of each state in order to freeze or reduce and, if possible, eliminate these undeclared programs in parallel with the reduction programs of the nuclear powers.

The committee's analysis does not assume a fundamental change in the nature of international relations in order to achieve these low levels of nuclear arms. It does assume unprecedented cooperation and transparency among all classes of nuclear powers on the specific issue of nuclear arms reductions. A few hundred nuclear weapons would be sufficient to deter nuclear attack through their potential to destroy essential elements of the society of any possible attacker. These remaining nuclear forces would have to be survivable and their command-and-control structure adequately redundant and robust; and widespread and effective national ballistic missile defenses must be absent. The operational posture of the much smaller forces must be designed for deliberate response rather than reaction in a matter of minutes.

PROHIBITION OF NUCLEAR WEAPONS?

The end of the Cold War has created conditions that open the possibility for serious consideration of proposals to prohibit the possession of nuclear weapons. It is not clear today how or when this could be achieved; what is clear is that comprehensive nuclear disarmament should be undertaken only in circumstances such that, on balance, it would enhance the security of the United States and the rest of the world.

The committee uses the word "prohibit" rather than "eliminate" or "abolish" because the world can never truly be free from the potential reappearance of nuclear weapons and their effects on international politics. Even the most effective verification system that can be envisioned would not produce complete confidence that a small number of nuclear weapons had not been hidden or fabricated in secret. More fundamentally, the knowledge of how to build nuclear weapons cannot be erased from the human mind. Even if every nuclear warhead were destroyed, the current nuclear weapons states, and a growing number of other technologically advanced states, would be able to build nuclear weapons within a few months or few years of a national decision to do so.

A durable prohibition on nuclear weapons would have three main benefits:

- It would virtually eliminate the possibility of use—whether authorized and deliberate or not—of nuclear weapons by states now possessing them. Viewed in light of the possibility of reconstitution of such arsenals in a crisis, prohibition can be seen as extending the dealerting measures recommended in the near-term part of the program—that is, increasing the time required to ready nuclear weapons for use from hours or days to months or years.
- It would reduce the likelihood that additional states will acquire nuclear weapons. Although the Nuclear Nonproliferation Treaty currently enjoys almost universal adherence, the nuclear weapons states cannot be confident of maintaining indefinitely a regime in which they proclaim nuclear weapons essential to their security while denying all others the right to possess them.
- It would deal decisively with the uncertain moral and legal status of nuclear weapons, as underlined by the recent advisory opinion of the International Court of Justice.

Nuclear disarmament poses risks as well as benefits, however:

- The prohibition on nuclear weapons might break down via cheating or overt withdrawal from the disarmament regime. To reduce these risks, a disarmament regime would have to be built within a larger international security system that would be capable not only of deterring or punishing the acquisition or use of nuclear weapons but also of responding to major aggression.
- Comprehensive nuclear disarmament could remove the moderating effect that nuclear weapons appear to have had on the behavior of states. The nuclear era represents the longest period without war among the major powers since the emergence of the modern nation state in the sixteenth century. Thus, it is argued that, if the major powers believed the risk of

nuclear war had been eliminated, they might initiate or intensify conflicts that might otherwise have been avoided or limited. But there have been, and continue to be, profound changes in the structure of the international order that are acting to reduce the probability of major war independent of nuclear deterrence. Moreover, even if all nuclear weapons were eliminated, the inherent capacities to rebuild them could act as a deterrent to the outbreak of major wars.

If the preconditions for agreed prohibition of nuclear weapons are met, however, the committee believes that a path to eventual prohibition can be found. One possible path for managing the transition to comprehensive nuclear disarmament would involve having an international agency assume joint or full custody of the arsenals remaining during the transition to prohibition. Alternatively, nations might find it preferable to bypass the intermediate step involving an international agency and proceed directly to negotiations to prohibit nuclear weapons either globally in a single agreement or in steps involving successive expansions in the number and geographical scope of nuclear weapon free zones.

It will not be easy to achieve the conditions necessary to make a durable global prohibition on the possession of nuclear weapons both desirable and feasible. Complete nuclear disarmament will require continued evolution of the international system toward collective action, transparency, and the rule of law; a comprehensive system of verification, which itself will require an unprecedented degree of cooperation and transparency; and safeguards to protect against the possibility of cheating or rapid breakout. As difficult as this may seem today, the process of reducing national nuclear arsenals to a few hundred warheads would lay much of the necessary groundwork. For example, the stringent verification requirements of an agreement on very low levels of nuclear weapons and fissile materials might by then have led to some new or expanded international agency with vigorous powers of inspection. The committee has concluded that the potential benefits of a global prohibition of nuclear weapons are so attractive relative to the attendant risks that increased attention is now warranted to studying and fostering the conditions that would have to be met to make prohibition desirable and feasible.

In any case, the regime of progressive constraints constituting the committee's proposed near- to midterm program makes good sense in its own right— as a prescription for reducing nuclear dangers without adverse impact on other U.S. security interests—regardless of one's view of the desirability and feasibility of ultimately moving to prohibition.

1

Why Change U.S. Nuclear
Weapons Policy?

The debate about appropriate purposes and policies for U.S. nuclear weapons has been under way for more than half a century, since the beginning of the nuclear age.[1] With the end of the Cold War, however, the debate about the roles of U.S. nuclear weapons (and those of other countries) has entered a new phase, propelled by the transformation of the international political landscape and the altered foreign policy challenges and opportunities that these changes are bringing about.

The committee's first major study of nuclear weapons policy appeared in 1991, early in the transition out of the Cold War.[2] Numerous studies exploring or proposing changes in nuclear forces and policies have been conducted since. The U.S. Nuclear Posture Review (NPR) completed in 1994 by the Department of Defense has been the basis of current U.S. nuclear policy.[3] Among unofficial studies, the series of reports on nuclear weapons issues by the Henry L. Stimson Center and the international studies of the prospects for the elimination of nuclear weapons conducted by the Pugwash Conferences on Science and World Affairs and by the Canberra Commission on the Elimination of Nuclear Weapons have attracted particular notice.[4] With the benefit of those studies and others, and the committee's own intensive re-examination of these matters—including wide-ranging discussions of nuclear weapons issues in the continuation of its long-standing series of meetings with counterpart groups in Russia, China, and Europe—the committee now offers a new assessment of the implications of the end of the Cold War and breakup of the Soviet Union for the future of U.S. nuclear weapons policy.

The remainder of this chapter first summarizes the reasons for believing that further changes in U.S. and Russian nuclear weapons policies are desirable and the prospects for near-term action on the two sides in this direction. Some dilemmas of nuclear deterrence that were instrumental in shaping nuclear arsenals during the Cold War, along with the dangers that resulted, are then briefly elaborated. The committee outlines its case for a regime of progressive constraints to continue the progress that has been made in reducing those dangers since the Cold War ended. The chapter concludes with an orientation to the remainder of the report and a caution about the economics of nuclear arms reductions.

THE PROBLEM AND THE PROSPECTS IN SUMMARY

Nuclear Weapons During and After the Cold War

During the Cold War, nuclear weapons were at the center of U.S. and Soviet national security strategies. Both countries developed large, diverse, dispersed, and accurate nuclear forces that were maintained at high alert levels. (Appendix B portrays the Cold War growth of nuclear forces, as well as the beginnings of their post-Cold War decline.) The officially stated rationales for these forces, in broad terms, were on the U.S. side to deter the Soviet Union from attacking or threatening to attack the United States or its allies with either conventional forces or nuclear weapons (see Box 1.1) and on the Soviet side to deny the United States and its allies any military or political advantage from their possession of nuclear weapons, and to be able to deliver a "crushing rebuff" to any use of nuclear weapons against the Soviet Union.

The actual events of the Cold War period are consistent with the view that the nuclear forces and policies of the two sides were successful in their stated purposes: from 1946 onward, neither side succeeded in consistently imposing its will on the other, neither waged major war against the other, and neither launched a nuclear attack against anyone. Of course, other factors were at work, including the memory of the vast destruction of World War II, nearly all of it accomplished with conventional forces. And proof of cause and effect is always elusive in international affairs, as is, even more generally, proof of why something did *not* happen.

But supposing that the nuclear forces and policies of the two sides were indeed major contributors to the avoidance of full-scale war between East and West in the post-1945 period, it still must be conceded that this outcome was accompanied by enormous risks. These risks included the danger that the nuclear arms competition might continue without limit, endlessly adding to destructive potentials, constantly risking some destabilizing imbalance, and forever tempting additional countries to acquire nuclear weaponry for purposes of protection, or status, in a world of nuclear-armed camps. Above all, the risks included the danger that an accidental, erroneous, or unauthorized launch of one or a few nuclear weapons, or some other escalatory dynamic arising out of political crisis or regional conflict, could lead to full-scale nuclear war and the unimaginable disaster that this would represent for civilization (see Box 1.2).[5]

BOX 1.1
Deterrence

Understanding the history of nuclear weapons policy and addressing the challenges of formulating new policies for the future both require an appreciation of the diverse definitions, applications, and dilemmas of deterrence. Because confusion can easily result from insufficient clarity about what is meant by the term in any particular context, the committee shall take pains to try to be clear—here and then throughout this report—about what it means by deterrence in the various forms and circumstances in which the concept has been applied.

The words "deter" and "deterrence" both derive from the Latin *deterrere*, to frighten from. The narrowest dictionary definition of "deter" in English, correspondingly, is "to discourage from some action by making the consequences seem frightening."* Both in everyday language and in the language of specialists in international politics and military strategy, however, "deter" has long since had a somewhat wider meaning: it is used not only to describe discouraging an action by the prospect of consequences that are frightening, but also for situations in which the restraint arises simply from the prospect of failure to achieve the intended aims, or the prospect of costs exceeding an action's expected benefits. Some writers in the literature of military strategy distinguish explicitly between deterring an attack by the threat of "punishment" (frightening consequences) and deterring an attack by the prospect of "denial" (of the objectives of the attack).**

"Deterrence" in the political/military context can refer either to measures taken to generate a credible prospect of punishment for an action, or of denial of its objectives, or of costs exceeding its benefits (i.e., the *practice* of deterrence) or to the state of restraint induced by such measures and by other factors (i.e., deterrence as a *condition*). Of course, how much is required in the way of the practice of deterrence to achieve an adequate condition of deterrence depends, among other things, on how attractive the aggressive act would be to its prospective perpetrator in the absence of deterrent measures and on how averse the prospective perpetrator is to punishment, failure, or unfavorable cost-benefit ratios. Certain intrinsic deterrent factors against aggressive acts in general, and against nuclear attack or threat of attack in particular, will generally be operative irrespective of any practice of deterrence. These factors in-

continued

---

*The New International Webster's Dictionary, (Naples, Fla: Trident Press International, 1995).

**See, for example, Lawrence Freedman, *The Evolution of Nuclear Strategy* (New York: St. Martin's Press, 1983).

BOX 1.1—continued

clude moral inhibitions, sense of kinship across national boundaries, and fear of loss of domestic political support. Although these are of variable importance from one case to the next and may not, in themselves, suffice to ensure restraint, the bigger they are in a given instance the less will be required of the practice of deterrence to augment them.

The practice of deterrence may entail creation of a credible prospect of retaliation in kind against the action that is to be deterred (e.g., the threat of invading an adversary's homeland if he invades yours) or of retaliation in a different (and possibly even nonmilitary) form (e.g., the threat of embargo of critical resources in response to an attack); it may entail erecting defenses, to decrease the chances of success of an attack and/or to increase the cost of a successful one; and often it will include a combination of these ingredients. Of course, the practice of deterrence has costs and risks as well as benefits: not only must deterrent measures be paid for, but they also may stimulate countermeasures by the putative adversary, or by others, that will necessitate still higher expenditures in the future if deterrence is to be maintained; the singling out of prospective adversaries and the brandishing of capabilities against them, which the practice of deterrence often entails, can aggravate tensions; and measures intended to enhance deterrence of premeditated attack (as, for example, by increasing the credibility of a retaliatory response) may increase the danger of war by inadvertence or accident.

This report is concerned mainly with *nuclear* deterrence, where "nuclear" refers to the character of the response that is contemplated, not necessarily to the kind of threat that is supposed to be deterred. In principle, nuclear deterrence could be used to deter not only nuclear attacks but also attacks with conventional forces, attacks with chemical or biological weapons, or even assaults on vital national interests by nonmilitary means.

Several terms for variants in the intended scope or mode of operation of nuclear deterrence are encountered widely enough in the nuclear weapons policy literature—and with enough variability and ambiguity in meaning—that it seems worthwhile to try to clarify them here. Specifically:

- The term *extended deterrence* has been used to mean extension of nuclear deterrence to deter not only attacks or coercion against the deterring country's own territory but also attacks or coercion against the territory of the deterring country's allies and also to mean deterrence not only of nuclear attacks/coercion but also of attacks/coercion based on conventional, chemical, or biological weapons. (The original U.S. conception of nuclear deterrence of the Soviet Union—even before that country had nuclear weapons— already contained both of these extended dimensions; explicit use

BOX 1.1—continued

of the adjective "extended" to describe what actions are to be deterred only came into use later, after some had proposed that nuclear deterrence should be restricted to deterring only nuclear attack or coercion and, in some arguments, only nuclear attack or coercion against one's own country.)

- The term *minimum deterrence* has been used in the literature with two different meanings. One meaning, referring to the scale of the contemplated nuclear response to the aggressive acts to be deterred, is that this scale is as small as possible consistent with still being sufficient to deter. The other meaning, referring to the range of threats to be deterred by the prospect of a nuclear response, is that nuclear deterrence relates only to threats of nuclear attack or, still more restrictively, only to threats of nuclear attack against the deterring country (and not, for example, to threats of nuclear attack against its allies).

- The term *existential deterrence* refers to a deterrent effect that arises from the mere existence of nuclear weapons in the possession of the countries or in the possession of their allies—or even from the existence of the capacity of a country or its allies to build nuclear weapons if they wished to do so—without any reliance on the "practice" of deterrence in the form of declared doctrines, specific weapons delivery capabilities, force postures, targeting plans, training exercises, or other actions intended to make it credible that carrying out the aggressive acts to be deterred would result in a nuclear response.

Finally, the *core function* of deterrence, or just *core deterrence*, in this report means the restricted form of extended nuclear deterrence in which coverage is intended against nuclear threats—and only nuclear threats—to one's own country and to one's allies. (This usage follows that in the committee's 1991 report, *The Future of the U.S.-Soviet Nuclear Relationship*. The committee does not use "minimum deterrence" for this purpose, although some have done so in the literature, because as indicated above that term also has had other meanings.)

Over the course of the Cold War, the two sides negotiated a series of arms control agreements to try to limit the direct dangers of their nuclear confrontation, and they were leaders in the construction of other agreements, with wider participation, intended to restrain the proliferation of nuclear weapons to additional countries. As with the argument "from history" for the success of nuclear deterrence itself, it can be argued that the facts are consistent with the view that these arms control and antiproliferation measures succeeded. There were no accidental, erroneous, or unauthorized launches of nuclear weapons and no escalation to nuclear war from a regional crisis—although crises there were. The nuclear arsenals were eventually capped, albeit at the very high levels of 30,000 to 40,000 nuclear weapons each for the United States and the Soviet Union.[6] The spread of nuclear weapons into the possession of additional countries proceeded much more slowly than most analysts of these matters had foreseen in the 1950s and 1960s. Now, with the impetus of the end of the Cold War, the nuclear arsenals of the United States and Russia are shrinking significantly, with physical dismantlement of warheads proceeding at a pace of 1,500 to 2,000 per year on each side.

A closer look at what happened in some of the crises of the Cold War, however, leaves room to question whether good fortune was not as much a factor as good management in avoiding escalation to disaster on these occasions.[7] A look at the sizes, compositions, and postures of the nuclear forces that remain today— and at those that will remain after the arms reduction agreements and unilateral decisions of recent years have been fully carried out—also gives cause for concern; such forces continue to be more formidable and more dangerous than necessary or appropriate for the conditions of the post-Cold War world.

The committee has concluded that, under the new circumstances, the security of the United States could be considerably enhanced by undertaking further reductions of nuclear forces globally, with accompanying changes in nuclear weapons policies and operational practices. The committee's conclusions and recommendations to this effect are based on a number of specific propositions— summarized for reference here and elaborated on in the remainder of this chapter and in Chapters 2 and 3—about the roles and dangers of nuclear weapons in the post-Cold War world.

Changing Roles, Circumstances, and Opportunities

The principal roles generally attributed to U.S. possession of nuclear weapons are (1) deterrence of premeditated nuclear attack; (2) deterrence of major conventional war; and (3) compensation for possible inadequacies in nonnuclear forces, including for deterrence or response to attacks with chemical or biological weapons. But these roles are less demanding or less relevant in the post-Cold War world than before, because of the following:

- The likelihood of all-out war between the United States and Russia has drastically diminished, and therefore the role of nuclear weapons can be narrowed significantly.
- The relative importance of regional conflicts has increased in the aftermath of the Cold War, but for conflicts of this type the practice of nuclear deterrence by the United States or Russia or the other declared nuclear weapons powers is likely to be unnecessary, irrelevant, ineffective, or even harmful in some cases.[8]

BOX 1.2
Accidental, Erroneous, or Unauthorized
Use of Nuclear Weapons

During the Cold War, U.S. nuclear policies were oriented toward deterring a deliberate, premeditated attack on the United States or its allies. The risk of a premeditated attack authorized by national leaders has diminished greatly with the end of the Cold War, but the risk of other kinds of nuclear attack—accidental, erroneous, or unauthorized—has not gone down proportionately.

Although the term "accidental" sometimes has been used in this context to cover a wide array of unintended and ill-considered actions, the committee uses it here in the narrower sense of the term "accident" to connote such events as programming mistakes or mechanical or electrical failures. The United States and Russia, and presumably the other nuclear weapons states as well, have worked very hard to ensure that nuclear weapons could not be launched or detonated as a result of equipment failures or operator errors. The 1994 detargeting initiative by Presidents Clinton and Yeltsin addressed this risk directly by agreeing that the United States and Russian would not target their missiles against each other on a day-to-day basis. While one cannot prove that such accidents are impossible, this type of risk probably is less worrisome than others that attend nuclear arsenals.

More serious, the committee believes, is the risk of erroneous use of nuclear weapons. Unlike accidents, an erroneous use of nuclear weapons would result from conscious decisions by military or political leaders, but these decisions would be based on incomplete or inaccurate information, faulty reasoning, misinterpretation of the intentions of other countries, and careless or hasty decisionmaking, perhaps influenced by the unintended consequences of prior actions. Possible examples include a decision to launch nuclear weapons in response to false or ambiguous warning of actual or impending attack, or misinterpreting a demonstration shot, unauthorized attack, or an attack on another country as a massive attack on one's own country. The reported deterioration of Russia's missile attack warning system is particularly troubling in this regard.

Another disturbing possibility is the theft or unauthorized use of nuclear weapons. Although no nuclear weapons state is immune to such risks, the general decline of morale in the Russian military is cause for special concern. It is generally believed that Russian nuclear weapons are accorded high levels of protection and security, but a further degeneration of the economy, domestic politics, relations with neighboring states, or civilian control over the military could dramatically increase the chance that a group, either inside or outside the military, might try to steal, use, or threaten to use nuclear weapons.

- The capabilities of U.S. conventional forces are formidable both in absolute terms and relative to the forces of potential adversaries and, with appropriate policies and allocation of adequate resources, will remain so, making it possible for the United States to respond effectively with conventional forces across a wide spectrum of threats, including attacks on the United States or its allies using chemical or biological weapons.

The principal dangers usually ascribed to U.S. possession of nuclear weapons (and their possession by others) are (1) nuclear war by accidental, erroneous, or unauthorized use of nuclear weapons, or by unintended escalation; (2) the failure of arms control, leading to excessive force levels; and (3) encouragement of nuclear proliferation. These dangers have not been shrinking, in the post-Cold War world, as rapidly as the relevance of nuclear weapons to U.S. security needs. Specifically:

- The post-Cold War changes in the sizes, compositions, and postures of U.S. and Russian nuclear forces—and in the doctrines governing the purposes and potential uses of these forces—have not kept pace with the changes in the post-Cold War military and political landscape. As a result, the risks of accidental, erroneous, or unauthorized use of nuclear weapons—which could then all too easily escalate—remain unacceptably high (possibly, on the Russian side, even higher than Cold War levels because of deterioration of the military and internal-security infrastructure and of morale).
- On both sides, the continuing competitive and even confrontational assumptions underlying some official discussions of the U.S.-Russian nuclear relationship, when coupled with the postures of the forces and the potential for destabilizing deployments of ballistic missile defenses, pose the risk that the arms control fabric woven during the Cold War and immediately thereafter could in fact unravel.
- Although the practice of nuclear deterrence by the United States, Russia, and the other declared nuclear weapons states can, in some instances, help inhibit the proliferation of nuclear weapons (by reassuring allies that they will be protected without needing to acquire their own nuclear weapons), in other circumstances the practice of nuclear deterrence is likely to aggravate proliferation dangers (by causing nonallies to feel threatened, by lending respectability to reliance on nuclear deterrence, and by undermining the credibility of the nuclear weapons states in their opposition to proliferation). The committee judges it likely, although it cannot be rigorously proved, that in the post-Cold War world the proliferation-aggravating effects of the practice of deterrence by the declared nuclear weapons states will increasingly outweigh the proliferation-inhibiting effects.
- Ultimately, it would prove difficult for the United States—the world's most powerful nation in conventional armaments—to continue to main-

tain that its security requires the possession of a strong nuclear deterrent while denying the validity of that argument for other nations.

These post-Cold War conditions and the likely long-term trends mean, in the committee's view, that the conspicuous role given to nuclear weapons during the Cold War can be greatly reduced *without significant adverse effect* on the probability of all-out war or on this country's capability to cope effectively with regional conflicts where its interests are at stake, and *with significant security benefits* in terms of improvements for relations with Russia and for the cause of nonproliferation and in terms of reduced risks of erroneous or unauthorized nuclear-weapon use and of inadvertent escalation. In addition, the trends that have made such changes possible and desirable are likely enough to continue that serious study of further longer-term changes in U.S. nuclear weapons policy is warranted.

The committee has concluded, accordingly, that the United States should pursue a two-part program of change in its nuclear weapons policies. The first part of the program is a near- to midterm set of mutual force reductions—together with accompanying changes in nuclear operations and declaratory policies and with measures to improve the security of nuclear weapons and fissile materials worldwide—to diminish further confrontational and potentially destabilizing aspects of force postures, to reduce the risks of erroneous, unauthorized, or accidental nuclear-weapons use, and to help curb the threat of further nuclear proliferation. The second part of the program is a long-term effort to foster international conditions in which the possession of nuclear weapons would no longer be seen as necessary or legitimate for the preservation of national and global security.

Prospects for Progress on the U.S. and Russian Sides

In its early phases the first part of the program is largely a bilateral U.S.-Russian one, and cooperation between the two countries is essential to its success. Remarkable progress in this bilateral effort has already been made. A decade ago few could have imagined the scope of the agreements that have been reached or are being considered to cut the nuclear arsenals of the United States and the former Soviet Union (see Box 1.3). The approach to further arms control taken in the NPR in 1994, however, was not sufficiently ambitious. The NPR's mandate was to rethink all aspects of U.S. nuclear weapons policy, and it did resolve some pressing force structure questions; but, in the end, it did not go very far toward addressing the most fundamental issues about appropriate numbers, roles, and postures of U.S. nuclear forces in light of the changes brought by the end of the Cold War. Specifically, the NPR did not recommend reductions in strategic forces beyond those already agreed to in the second Strategic Arms Reduction Treaty (START II), and it did not alter the "weapons of last resort" mission for U.S. nuclear forces that allows their first use in response to non-

BOX 1.3

U.S. and Soviet/Russian Accomplishments in Nuclear Arms Reductions and Operational Arms Control

1987 *Intermediate-Range Nuclear Forces (INF) Treaty*
Bans all U.S. and Soviet ground-launched ballistic and cruise missiles with ranges between 500 and 5,500 kilometers.

1990 *Last Soviet nuclear weapons test*

1991 *Strategic Arms Reduction Treaty (START) I signed*
Reduces the number of deployed strategic warheads from about 11,000 for Russia and 13,000 for the United States to about 8,000 on each side.

1991 *U.S. and Soviet/Russian unilateral initiatives to reduce*
-92 *nonstrategic weapons*
All U.S. ground-launched and sea-launched nonstrategic nuclear weapons to be withdrawn to the United States and all Soviet/ Russian nonstrategic nuclear weapons to be withdrawn to Russia. Thousands of U.S. and Russian weapons to be destroyed.

1991 *U.S. takes all strategic bombers and "Looking Glass" airborne command posts off alert*

1992 *Last U.S. nuclear weapons test*

1992 *U.S. and Russia begin Cooperative Threat Reduction Program with funding from the Nunn-Lugar Act*

1992 *Lisbon Protocol*
Ukraine, Kazakstan, and Belarus agree to adhere to START I as nonnuclear weapons states and to return all nuclear weapons on their territories to Russia.

1993 *U.S. and Russia sign agreement for the disposition of excess highly-enriched uranium (HEU) from dismantled Soviet nuclear weapons*
Low-enriched reactor fuel derived from 500 tons of Russian weapons-usable HEU to be shipped to the United States for sale on the world market.

BOX 1.3—*continued*

1993 *START II signed*
Limits the number of deployed strategic warheads to 3,000 to 3,500 on each side. Ratified by the United States in January 1996 but not yet ratified by Russia.

1994 *START I enters into force*
With formal adherence of Ukraine, Kazakstan, and Belarus to the NPT as nonnuclear weapons states.

1994 *U.S.-Russian summit agreement not to target nuclear missiles on one another*

1995 *Indefinite extension of the NPT*

1996 *Comprehensive Test Ban Treaty signed by the five nuclear weapons states*

1997 *U.S.-Russian summit agreement to begin START III negotiations upon entry into force of START II*
Agreement in principle to limit each side to no more than 2,000 to 2,500 deployed strategic warheads.

nuclear attacks.[9] (Past and current U.S. positions on these and other aspects of nuclear policy are discussed in detail in Chapter 2.)

The March 1997 Helsinki summit agreement to seek a START III treaty with a level of 2,000 to 2,500 deployed strategic warheads on each side is a welcome sign of the importance the U.S. and Russian presidents attach to continuing reductions in the two countries' strategic forces.[10] What remains to be seen is how quickly other influential officials and institutions in the two countries will go along.

With respect to Russia, there are reasons for concern that progress will not be easy. The post-Cold War diminution in Russian military power, particularly the weakening of its conventional armed forces made evident by their poor performance in Chechnya, have led many Russian military and political leaders to reemphasize the importance of nuclear weapons.[11]

Some in Moscow see nuclear weapons as providing specific policy options against a range of purported threats, as the response to the prospective enlargement of NATO has made clear. Commentators—including high-level figures such as former Minister of Defense Rodionov—have stressed that, if NATO moves toward Russian borders, Russia might move nuclear weapons westward. This emphasis on the military utility and policy relevance of nuclear weapons, if maintained at a high rhetorical pitch, may make it difficult for Russian policy-makers to pursue further deep reductions, unless the reductions can be portrayed as "correcting" inequalities of past agreements—as the Russians argue that START III should "correct" START II. The utility and relevance arguments also may make it difficult for Russian leaders to pursue the goal of nuclear disarmament.

U.S. leaders must take this developing Russian perspective seriously in designing future U.S. nuclear weapons policy, since most if not all of the desirable adjustments require corresponding Russian actions. Conditioning START III negotiations on entry into force of START II gives substantial leverage to the Duma, where opposition to arms reductions is strong.

In addition, the severe funding shortages plaguing Russian government operations may limit Russia's ability to bear the costs of nuclear force reductions. Despite the progress made to date, funding for further reductions may be hard to justify domestically if soldiers' wages are inadequate or unpaid. Cooperation with the West for nuclear security and the implementation of force reductions has been helpful and these efforts, in which the United States has played a leading role since 1991 through the Nunn-Lugar Cooperative Threat Reduction Program, complement other U.S. nuclear policy initiatives. The scope of the problem—particularly improved security for fissile materials—is larger than can be addressed by the cooperative efforts undertaken thus far. It is very much in the West's own security interests to continue this nuclear cooperation, but ultimately, the broader dilemma must be addressed by Russia. Because this committee and other National Research Council committees have offered detailed analysis and recommendations on these problems in other reports, this report concentrates on other nuclear policy issues.[12]

Fortunately, several ways to address the military problems perceived by the Russians are available, some dependent on Russian actions alone and others requiring action on a reciprocal (and in some cases negotiated) basis. The Russians themselves, for example, have recognized the need to downsize and modernize their conventional forces. If this military reform effort can be gotten under way, it will address the weakness of the conventional forces from within. Adaptation of the Conventional Forces in Europe Treaty is another path that has been opened to address Russian concerns, including, perhaps, those arising from the prospective enlargement of NATO.

On the nuclear side, timely and flexible pursuit of further strategic arms reductions should help overcome the obstacles to Russian ratification of START II (about which more is said in Chapters 2 and 3). Attention to nonstrategic and reserve nuclear weapons in these further discussions should also be helpful.[13] If the United States and Russia, in full cooperation with NATO, were able to agree to foreclose forward deployments of nuclear weapons in Europe in a mutual, reciprocal, and verifiable manner, this step should contribute to deemphasizing the role of nuclear weapons in Europe. It would be a clear signal that both Russia and NATO are committed once and for all to denuclearization of the former East-West confrontation.

The Wider and Longer-Term Issues

It is essential that the near-term program for nuclear weapons policy also address the critical global problem posed by the continued risk of further nuclear proliferation. The effort to prevent the spread of nuclear weapons must employ new approaches and a determined international consensus. Deterrence practiced by two powerful and opposing alliance systems was dangerous enough. Multiple confrontations among the dozens of states that could acquire nuclear weapons, if they chose to do so, could prove unmanageable. If dozens of states do acquire nuclear weapons, it will increase the risk that terrorists or even criminal organizations may obtain them as well. Strengthening the consensus against nuclear proliferation, finding ways to engage the three undeclared nuclear weapons states in arms reductions, increasing the security of nuclear weapons and fissile materials worldwide, and dissuading those few countries still bent on acquiring nuclear weapons must be top U.S. priorities.

As for the second, long-term part of the program, the United States is committed through its adherence to the Treaty on the Nonproliferation of Nuclear Weapons (referred to hereinafter as the Nonproliferation Treaty or NPT) to pursue the goal of eventual elimination of nuclear weapons and, along with the other declared nuclear weapons states, reaffirmed that commitment at the 1995 NPT Review and Extension Conference. The committee recognizes that fundamental modifications of international political relationships not now foreseeable will be required to achieve such a goal. Surmounting the multilateral challenge posed by the near-term goal of deep force reductions and changes in nuclear operations, however, will contribute to progress on this second part of the program by demonstrating that a cooperative effort is feasible and by developing the technical monitoring and verification systems required.

NUCLEAR WEAPON DILEMMAS AND DYNAMICS

The unprecedented destructive power of nuclear weapons fundamentally changed the offense-defense balance in military conflict, since even a single large nuclear warhead that managed to penetrate deployed defenses could destroy a great city. As a result, the standards that defensive measures must meet in order to defend a nation against nuclear weapons are much higher than those sufficient for defense against more traditional military threats, and in fact are unlikely to be attainable against significant offensive forces equipped with countermeasures.

Delivery means for nuclear weapons are diverse, and, to be effective, national defenses must protect against all of them. This means that attempts to defend the United States or its allies against nuclear attacks on their populations could be overcome with much less effort than would have to be invested in the defenses. "Offense dominance" thus prevailed from the time that the Soviet Union acquired the capacity to deliver nuclear weapons with intercontinental ballistic missiles, and in the committee's judgment no subsequent developments have altered that basic situation.

In an era in which the hostility of the U.S.-Soviet relationship made it seem imprudent to rely on good intentions to preclude nuclear attack and in which, early on, invulnerable basing modes for significant parts of each side's nuclear forces made it impractical to execute a disarming first strike if conflict seemed imminent, the apparently inescapable impotence of defense reinforced inclinations to rely on deterrence through the threat of retaliation. No more effective alternative was apparent.

Analysts and political leaders alike soon came to recognize, however, that nuclear deterrence itself was (and is) burdened with an array of contradictions and dilemmas. For example, deterrence is only likely to succeed if there are credible plans for what to do if it fails, but constructing such plans is exceedingly difficult. More specifically, for deterrence to work, the prospective attacker must believe that the threat to retaliate might actually be executed. To increase the credibility of the response, the deterrer constructs a war plan (and the nuclear forces to support it) in which, after suffering an initial attack, the deterrer would gain relative to the undamaged attacker. But this condition is very difficult to satisfy, especially if one assumes that the initial attacker has held back some nuclear weapons for a further strike against the retaliator.

Each side in such a confrontation is motivated to try to shore up the credibility of its nuclear deterrent threat by decreasing the vulnerability of its retaliatory forces to a first strike while increasing its capacity to destroy, in a second strike, the remaining nuclear forces of the initial attacker (in order to limit the damage from a counterretaliation). But this simultaneous pursuit of invulnerability of one's own nuclear forces and counterforce capabilities against the nuclear forces of one's adversary is likely to look, to the adversary, like an attempt to gain the capability to carry out a *first* nuclear strike with relative impunity. Neither side would want to allow the other to achieve such a first-strike capability (or to suffer the delusion that it had achieved it) since, in the kind of hostile relationship that gives rise to the practice of nuclear deterrence in the first place, even the impression that there is a decisive advantage to striking first is extremely dangerous, especially during a crisis.

A theoretical alternative to the invulnerability/counterforce approach to shoring up the credibility of the threat to retaliate is to arrange things so that the retaliation would be *automatic*, as, for example, an arrangement to launch all or part of one's retaliatory nuclear forces automatically upon receipt of electronic warning that an adversary has launched a nuclear attack. Such a launch-on-warning posture, if an adversary believed it had actually been implemented, certainly would increase the credibility of the retaliatory threat. But the increase in credibility would come at the cost of an increased chance of initiating nuclear holocaust accidently, as a result of false warning or other malfunction in the automated command-and-control system. Survivable basing combined with delayed automatic response would remove some, but not all, of these problems.

The dilemmas at the heart of nuclear deterrence, then, arise in part because attempts to make the threat of retaliation credible can be seen as aggressive advantage seeking by the other side. This raises tensions and stimulates countervailing measures, hence arms races, or increases the chance of nuclear war from crisis instability or accident.

There is also the dilemma of deciding "How much is enough?" in the sense of how many nuclear weapons, of what destructive power, delivered with what degree of assurance, against what set of targets will suffice to deter a country's potential adversaries, in all the diversity and unknowability of their motivation and mental state. There is the related dilemma of whether an amount judged to be enough, for purposes of making the retaliatory threat, would ever be seen as proportionate or appropriate by the leaders who have to decide whether to carry out the threat after deterrence has failed. The dilemmas of secrecy, wherein the adversary needs to know something of the plans for retaliation in order to be deterred, but must not know too much lest this enable him to take countermeasures that would reduce the retaliation's effectiveness—and, hence, the effectiveness of the threat—also arise. Finally, there is the dilemma, discussed earlier, that the assertion by some countries of a need and right to practice nuclear deterrence may eventually encourage additional countries to assert the same need and right, leading to proliferation of nuclear weapons and, hence, a more dangerous world.

These inherent dilemmas of the practice of nuclear deterrence were compounded, during the Cold War, by the U.S. threat of first use of nuclear weapons against a Soviet conventional attack in Europe. Because of the difficulty of convincing both the Soviet Union and U.S. allies that the United States would really use nuclear weapons in this circumstance, and because of the difficulty of devising reasonable targeting strategies for this eventuality given the large-scale destructive potential of nuclear weapons, the U.S. search for ways to shore up the credibility of its nuclear deterrent was even more energetic and wide ranging than it would otherwise have been. In the beginning this aim was accomplished through a massive superiority in strategic weapons, which gave the United States a reasonable prospect of a successful first strike. This strategic superiority was soon augmented by U.S. deployment of thousands of nonstrategic ("tactical") nuclear weapons in Europe. As the Soviet Union attained a survivable nuclear force in the 1960s, however, the threat of a massive U.S. response to limited Soviet attacks was no longer credible. The United States responded by bolstering its conventional defenses and those of its allies, by diversifying and expanding its nonstrategic nuclear deployments and developing limited and selective nuclear options, and by deploying thousands of strategic warheads of greater accuracy to improve the prospects for successful counterforce attacks.

The expansion of the U.S. and Soviet nuclear arsenals to tens of thousands of warheads was thus not a consequence of a carefully thought out, fully understood process based on an overarching nuclear doctrine. Although each side articulated the overall purposes that it intended nuclear weapons to serve, details of the nuclear doctrines that ostensibly governed planning and operations for the deployed nuclear forces often followed nuclear deployments rather than shaped them. The size and composition of the two sides' forces were driven by the interactive dynamic generated by the dilemmas of deterrence as just outlined, frequently controlled by technological and financial limitations and opportunities, but further complicated by the interactions of domestic and international politics and by tendencies toward worst-case assessment in the face of uncertainties about the capabilities and intentions of the other side. The result was a nuclear arms race in which the numbers, sophistication, and alert levels of U.S. and Soviet nuclear forces grew to levels difficult to comprehend by anyone other than those who were involved in the process—and often incomprehensible even to them.[14]

THE CASE FOR POST-COLD WAR REDUCTIONS
AND TRANSFORMATIONS

This committee has concluded that the continuing dilemmas and dangers of nuclear deterrence as practiced in the past by the United States can and should be alleviated in the post-Cold War security environment by confining such deterrence to the *core function* of deterring nuclear attack, or coercion by threat of nuclear attack, against the United States or its allies. That is, the United States would not threaten to respond with nuclear weapons against conventional, chemical, or biological attacks. The committee believes that Russia and the other nuclear weapons states can be persuaded to reach a corresponding conclusion.

With regard to chemical and biological weapons in particular, as discussed in Chapter 2, the committee has determined that their indiscriminate and often unpredictable effects, as well as the potential for defenses against them, make CBW weapons of restricted utility in achieving strategic advantage in war. If chemical or biological agents were used as weapons of terror by state or non-state actors, the committee concludes that nuclear weapons would be ineffective as a deterrent in advance of such an attack because they would be recognizably difficult to deliver in a timely and targeted manner against the perpetrator. Precise and technically capable conventional weapons could effectively provide the response and would avoid the broader consequences of nuclear use.

In all likelihood the United States will consider it necessary to continue to rely on the core function of deterrence as long as nuclear weapons continue to exist in the possession of states that conceivably might consider using them against this country or its allies. The committee assumes that some—although it is hoped not all—other nuclear weapons states will similarly consider it necessary to retain some nuclear weapons for "core deterrence." The size and scope of the efforts deemed necessary to fulfill the core function, however, presumably will shrink in parallel with what the committee hopes is the declining plausibility, over time, that any state would consider mounting a nuclear attack against anyone. There are strong reasons to make every effort to hasten the arrival of international conditions in which threats of nuclear attack are simply no longer thinkable. Under such conditions, the practice of deterrence with all its dilemmas and dangers would no longer be necessary.

As long as nuclear weapons exist, this very existence will exert a deterrent effect—*existential deterrence*—against unrestricted conventional wars among the major powers, since it will be recognized that such conflict in a world with nuclear weapons might well lead to their use, with intolerable destruction as the result. Indeed, even the existence of the *idea* of nuclear weapons—more specifically, the ability of many states to make them—is enough to create an existential deterrent effect against large-scale conflicts of all kinds. That is not to say that this effect would necessarily always be sufficient to prevent conflict in the future, as it has not always been in the past. But it could provide *part* of the assurance required, in an international system much different than today's, that all-out wars of any kind are unlikely to occur.

If, in the meantime, only the core function of nuclear weapons retains validity, fundamental changes in the nuclear force structures and operational practices of the major nuclear powers become both possible and desirable. The core function can be performed by far smaller nuclear forces than the United States now deploys, provided that these forces are survivable and can reach their targets.

There is also both symbolic and political value in having the United States actively pursue further reductions and changes in nuclear operations. As the first country to develop nuclear weapons, the only country to explode them in war, and the country that has consistently taken the lead in efforts to control them, the United States has a unique interest and irreplaceable role in reinforcing the norm that nuclear weapons will not be used for coercive purposes.

Benefits of the Proposed Changes

Nuclear force reductions and certain changes in nuclear operations would increase U.S. and global security in important ways.

First, reducing U.S. and Russian nuclear forces and revising operations for the mission of fulfilling only the core function will decrease the continuing risk of accidental, erroneous, or unauthorized use of nuclear weapons for several reasons. Smaller arsenals will be easier to safeguard and protect from accident, theft, and unauthorized use, not only by virtue of reduced numbers of weapons to monitor at a smaller number of sites but also by permitting retention of only those weapons with the most modern safety and security features. Reducing alert rates, decreasing capacities to use nuclear weapons quickly and with little warning, abandoning plans for the rapid use of nuclear weapons, and deploying cooperative measures to assure states that forces are not being readied for attack should reduce the probability and consequences of erroneous nuclear-weapons use—for example, on false warning of attack—particularly during a crisis. (Of course it is extremely important to take care that reductions in deployed nuclear warheads and dismantlement of the warheads made surplus as a result do not lead to countervailing increases in the dangers of theft and unauthorized use as a consequence of inattention to the challenges of safe storage of these weapons and the nuclear materials removed from them.)[15]

Second, further reductions will bolster the nuclear nonproliferation regime. If U.S. foreign and defense policy continues to rely heavily on nuclear weapons while attempting to deny others the right even to possess such weapons, the effectiveness of U.S. arguments against proliferation will be weakened. U.S. and Russian nuclear reductions will not in themselves dissuade a state bent on acquiring nuclear weapons; today's undeclared nuclear powers and would-be proliferators are driven above all by regional security concerns. In such cases, the denial of material and technical resources and a combination of political and economic incentives and disincentives would provide the greatest leverage. But U.S. and Russian progress in arms reductions helps shore up global support for anti-proliferation measures; and failure to make such progress can strengthen the influence of those arguing for nuclear weapons acquisition in countries where this is under internal debate.

Third, continued actions by the United States and Russia to reduce their nuclear arsenals—and the roles and missions assigned to the arsenals—will help induce the other declared and undeclared nuclear weapons states to join the arms control process. At the levels planned under the NPR, for example, under which it is estimated that the United States and Russia each would retain a total of about 10,000 nuclear warheads, deployed and in reserve, the other nuclear powers have little motivation to submit their much smaller arsenals to any form of control.

Is It Prudent to "Hedge"?

A central consideration in the NPR's conclusions was the perceived need to retain U.S. flexibility in the event of the reversal of reform in Russia. As a result, at present the United States has opted to maintain a "hedge" to provide the ability to reconstitute nuclear forces if the need arises.[16] Under START II, both sides would retain the capability in a crisis to deploy thousands of additional warheads by increasing warhead loadings on existing missiles and bombers. But in reality the United States has a far greater potential for uploading than Russia because of the technical capabilities of U.S. delivery vehicles.

The committee believes that the time has come to reconsider the need for such a hedge. Deploying yet more firepower in the event of renewed political antagonism with Russia would not improve the practical deterrent effect of U.S. nuclear forces. Moreover, the ability to overtly increase strategic readiness—by dispersing bombers and by moving a larger fraction of the ballistic missile submarine force to patrol areas—would provide a hedge against surprise. Increases in U.S. nuclear force levels would be necessary only if massive growth in the Russian force imperiled the survivability of the U.S. arsenal. For the foreseeable future Russia has no realistic capability to make such reconstitution possible.

The hedge strategy could become a self-fulfilling prophecy: the substantial stock of reserve warheads that the United States considers prudent could look to Russia very much like an institutionalized capability to break out of the START agreements. Russian legislators, worried about the breakout potential of U.S. forces and the high monetary cost of compliance, are already resisting the ratification of START II, which requires Russia to eliminate all of its multiple-warhead land-based ICBMs.[17] To the extent that the United States regards a return to hostile relations as a concern, it should focus on decreasing the probability of such developments.[18]

Creating a Regime of Progressive Constraints

In view of the foregoing, the committee believes that the United States and the other declared nuclear powers can preserve the core deterrent function of nuclear weapons with deployment levels substantially lower than those in current plans. In addition, substantial adjustments can be made in the operational practices governing existing nuclear weapons. The nuclear powers can also achieve higher standards of security for their nuclear warheads and for fissile materials worldwide.

The transformation of Cold War deterrent practices to adjust to new international security circumstances requires a balanced program to make many types of changes. The next two chapters develop the arguments for a revised set of U.S. policies that would lead to very low levels of nuclear forces and significantly reduced risk of their use. In addition to force reductions, key features of this proposed set of policies include:

- Renunciation of the first use of nuclear weapons for any purpose and the restriction of retaliation solely to attacks involving the use of nuclear weapons.

- Termination of alert practices in which weapons are deployed in configurations ready for immediate use (within a few minutes or hours). Any return of nuclear weapons to continuous-alert status would be renounced; preparing nuclear weapons for immediate delivery would be a result, as well as a signal, of serious military intent to use them.

- Emphasis on force structures that would guarantee the survivability of a significant portion of the remaining delivery vehicles.
- Termination of the previous emphasis on preparing for immediate, large-scale retaliation and mass targeting. Targeting preparations would be done as part of contingency planning under the presumption that any actual use of nuclear weapons would be specifically targeted in response to the circumstances at the time.
- The development and acceptance of high standards of physical protection and accounting for nuclear warheads and for all fissile materials.

ORIENTATION AND A CAUTION

The remainder of this report elaborates the arguments and findings summarized in this chapter. Chapter 2 reviews current U.S. nuclear weapons policy, U.S.-Russian nuclear relations, and the problem of global nuclear proliferation, laying the foundation for Chapter 3's recommendations for a regime of progressive constraints. These recommendations include proposals for the operational transformation described above and for successive U.S.-Russian force reductions. These reductions would begin with a quick cut to about 2,000 deployed strategic warheads each as envisioned in the Helsinki summit, then move to reductions to a total inventory of about 1,000 warheads each, and finally to a *total* inventory of a few hundred warheads on each side. Chapter 4 takes up the long-term agenda, exploring the conditions under which it might be possible to prohibit nuclear weapons altogether and the possible paths to reach that goal.

In drawing conclusions and making recommendations, the committee was motivated primarily by concerns about national security, international stability, and international obligations—and much less with economics. Advocacy for cuts in U.S. nuclear weaponry has often raised hopes that these reductions might also be a substantial factor in reducing the military budget. Recent studies of past expenditures by the U.S. nuclear weapons program, which include the cost of both the nuclear explosives and of the delivery systems, show how great the past costs have been.[19]

In the present study, however, the committee is *not* predicting that nuclear force reductions would save a great deal of money. The cost directly attributable to nuclear weapons would indeed decrease as the role of nuclear weapons is reduced. But there are additional costs associated with such a shrinkage, including the direct costs of dismantling and the verification and other costs associated with the arms control regime necessary to maintain compliance. Most important, however, is the question of the level of conventional forces that this country deems necessary in the face of a deemphasis on nuclear weapons. This is a matter well beyond the scope of this study, and the state of international relations over decisions about overall military requirements will have much more impact on future military budgets than the changes in nuclear weapons posture recommended here.

NOTES

1. The debate began among Manhattan Project scientists and, separately, in the top leadership of the U.S. government, even before the Trinity nuclear test in 1945. See, for example, Richard Rhodes, *The Making of the Atomic Bomb* (New York: Touchstone, 1988).

2. National Academy of Sciences, Committee on International Security and Arms Control, *The Future of the U.S.-Soviet Nuclear Relationship* (Washington, D.C.: National Academy Press, 1991).

3. The Nuclear Posture Review did not lead to a public report; its contents were communicated, however, through a widely disseminated set of unclassified briefing charts.

4. The Henry L. Stimson Center's reports include *Beyond the Nuclear Peril: The Year in Review and the Years Ahead* (1995), *An Evolving US Nuclear Posture* (1995), and *An American Legacy: Building a Nuclear-Weapon-Free World* (1997) (Washington, D.C.: The Henry L. Stimson Center). The principal Pugwash and Canberra reports are, respectively, J. Rotblat et al. (eds.), *A Nuclear Weapons Free World: Desirable? Feasible?* (Boulder, Colo.: Westview Press, 1993); and Canberra Commission on the Elimination of Nuclear Weapons, *Report of the Canberra Commission on the Elimination of Nuclear Weapons* (Canberra, Australia: Commonwealth of Australia, 1996).

5. The distinction among accidental, erroneous, and unauthorized uses of nuclear weapons is elaborated in Box 1.2.

6. See Appendix B.

7. See, for example, James G. Blight and David A. Welch, "Risking 'The Destruction of Nations': Lessons of the Cuban Missile Crisis for New and Aspiring Nuclear States," *Security Studies*, vol. 4, no. 4; Scott D. Sagan, "The Perils of Proliferation: Organization Theory, Deterrence Theory, and the Spread of Nuclear Weapons," *International Security*, vol. 18, no. 4 (Spring 1994), pp. 66-107; and Scott D. Sagan, *Moving Targets* (Princeton, N.J.: Princeton University Press, 1989).

8. By "practice of deterrence" the committee means the measures taken to generate a credible prospect of punishment for an action, or of denial of its objectives, or of its costs exceeding its benefits. See Box 1.1 for a further discussion.

9. This "last resort" formulation, which was promulgated during the Bush administration and formally adopted by NATO in 1990, is the current variant of the first-use-if-necessary policy on the employment of nuclear weapons that served as U.S. and NATO policy throughout the Cold War. See NATO Press Communique S-1(90)36, "London Declaration on a Transformed North Atlantic Alliance, Issued by the Heads of State and Government Participating in the Meeting of the North Atlantic Council in London on 5th-6th July 1990."

10. The White House, Office of the Press Secretary, "Fact Sheet: Joint Statement on Parameters on Further Reductions in Nuclear Forces," March 21, 1997.

11. See, for example, "Strategic forces now in forefront of Russia's defence—commander," BBC Summary of World Broadcasts, Part 1, The Former USSR, December 19, 1996, No. SU/2799, pp. S1/1-S1/2.

12. For recent assessments of some of the major U.S. cooperative programs in this area, see *Proliferation Concerns: Assessing U.S. Efforts to Help Contain Nuclear and Other Dangerous Materials and Technologies in the Former Soviet Union* (1997) and *An Assessment of the International Science and Technology Center* (1996). Two CISAC studies of the particular problem of excess weapons plutonium are *Management and Disposition of Excess Weapons Plutonium* (1994) and *Management and Disposition of Excess Weapons Plutonium: Reactor-Related Options* (1995). All were published by the National Academy Press.

13. See Box 2.1 for definitions of strategic, nonstrategic, and reserve weapons.

14. For example, see Herbert F. York, *Race to Oblivion* (New York: Simon and Schuster, 1970) and Robert S. McNamara with Brian VanDeMark, *In Retrospect* (New York: Time Books, 1995).

15. As noted above, the problems of safeguarding surplus nuclear warheads and nuclear materials are not treated in detail in this report, notwithstanding their great importance, because they were the focus of a separate, two-volume CISAC study as well as the NRC study *Proliferation Concerns*.

16. "A significant shift in the Russian government into the hands of arch-conservatives could

restore the strategic nuclear threat to the United States literally overnight. . . . The NPR called for an affordable hedge in which the approved force structure could support weapons levels greater that those called for under START should major geostrategic changes demand it." (William J. Perry, *Annual Report to the President and Congress*, U.S. Government Printing Office, Washington, D.C., February 1995, p. 86).

17. Russia is required to destroy its multiple-warhead SS-18s and SS-24s. It will retain many of its SS-19s, but they will be downloaded to carry only a single warhead.

18. Given the years required for a hypothetically hostile Russia to reconstitute conventional forces capable of challenging the United States and its allies, the time should be sufficient for compensating action.

19. For example, see Stephen J. Schwartz, "Four Trillion Dollars and Counting," *Bulletin of the Atomic Scientists*, Nov.-Dec. 1995, pp. 32-52, which summarizes the work of the U.S. Nuclear Weapons Cost Study project.

2

Current U.S. Nuclear Weapons Policy

This chapter's treatment of current U.S. nuclear weapons policy is divided into two parts. The first part, on the U.S.-Russian nuclear interaction, discusses the recent history of—and current problems and opportunities presented by—the nuclear weapons relationship between the two countries, including issues of deeper reductions, nuclear operations (alert levels and targeting), and ballistic missile defenses. The second part, on nuclear weapons policy and nonproliferation, covers the global nonproliferation regime, U.S. positive and negative security assurances and guarantees, and counterproliferation policy.

THE U.S.-RUSSIAN INTERACTION

Six years have elapsed since the breakup of the Soviet Union. Although many uncertainties remain about the role that Russia will assume in the world, the era of Soviet Communism has ended. The Russian presidential election in the summer of 1996 demonstrated that the Russian people have moved beyond the Soviet period. Russia could falter in its quest for constitutional democracy and instead become a nationalist dictatorship, an oligarchy, or a failed federal system. All of these would be serious setbacks for Russia and would pose varying degrees of risk to the security of the West. But Russia is not seeking to resume its position as the political and ideological leader of an anti-Western camp in what Marxist-Leninists used to call the clash of two opposing social systems.

Russia is no longer trying to develop a closed economic, political, and security system based on the superiority of its ideology and instead is attempting to transform itself into an open society. The Cold War world, characterized by a potentially violent East-West standoff, is being replaced by a world in which

Russia seeks to be an active member of the global economy. The fundamental economic imperatives driving Russian reform are especially important to creating the impetus for Russia to cooperate with the international community.

Although less certain, considerable military and political cooperation also may develop. The regional security challenges of the Middle East, South Asia, the Pacific Rim, and Europe no longer reflect the old Cold War context. Nuclear weapons still remain a key element of Russia's political and military status, but they will not determine its success or failure as an international actor. That will be defined mainly by the country's overall economic success.

Russia's desire for economic advancement is more clearly established than its political and strategic position. The prospect of NATO expansion into Eastern Europe has triggered deep-seated Russian concerns about its future security that have been easily exploited by extreme nationalists and antidemocratic groups. The rhetoric differs sharply, however, from the past seven decades, since it is no longer animated by a hostile ideology.

The United States and its allies have recognized the desirability of close engagement with Russia, despite the uncertainties inherent in its continuing transition. The reduction and destruction of the vast nuclear arsenals developed during the Cold War are among the most important domains of such collaboration. Substantial progress has already been made in adapting the nuclear forces of the United States and Russia to the post-Cold War environment. The first Strategic Arms Reduction Treaty (START I), the last Cold War arms agreement, was signed in 1991. It is now being implemented by both countries and will reduce the number of deployed strategic warheads from about 11,000 for Russia and 13,000 for the United States to about 8,000 on each side.[1] START II, signed in 1993 and ratified by the United States in early 1996 but (at this writing) not yet ratified by Russia, would further limit the actual number of deployed strategic warheads to 3,000 to 3,500 on each side (see Box 2.1). Through unilateral actions, the United States has reduced the number of its deployed nonstrategic warheads by 90 percent, from over 10,000 to about 1,000 warheads, all of which are bombs to be carried on dual-purpose aircraft. In reciprocal initiatives, Russia has made substantial (but less quantifiable) reductions in its nonstrategic warheads.

In addition to reducing their arsenals, both sides have undertaken a number of other measures. They have ended nuclear testing, and for the first time since World War II the United States is developing no new types of nuclear weapons or nuclear delivery systems.[2] The United States has taken all of its strategic bombers off alert and its airborne military command posts no longer fly continuous-alert missions. The United States and Russia have agreed not to target their missiles against each other on a day-to-day basis as a precaution against the consequences of an accidental launch. Production of weapons-grade fissile material has stopped in the United States and is continuing to a small extent in Russia only until such time as the reactor cores in three dual-purpose plutonium production reactors have been converted.[3]

BOX 2.1
Accounting for Warheads Under the START Treaties

The term "accountable" weapons refers to the number resulting from application of counting rules in START I, which are based on counting deployed delivery vehicles and assigning an agreed number of warheads to each type of vehicle. While the START I counting rules assume ICBMs and SLBMs to be armed with the maximum number of warheads with which the various missile types have been tested, these rules substantially undercount gravity bombs and short-range attack missiles (SRAMs) inasmuch as bombers not armed with cruise missiles are counted as if they contained only one warhead regardless of how many bombs and SRAMs they might actually carry.

The terms "active," "operational," or "deployed" as applied to nuclear weapons/warheads are synonymous; they all refer to the portion of a country's nuclear bombs and warheads that could be delivered by that country's operational delivery systems. In the present study the term "deployed" is used to denote this category. The term "inactive" refers to intact bombs and warheads beyond those that could be delivered as just indicated. These are often characterized as "spares" or "reserves." "Strategic" nuclear weapons/warheads are those intended for delivery on missiles or bombers with ranges over 5,500 kilometers; "nonstrategic" are those to be delivered by shorter-range systems.

START II allows the United States and Russia to keep 3,500 deployed "strategic" nuclear warheads each. START II does not limit nondeployed strategic warheads and the United States plans to keep up to 5,000 of them in various levels of readiness. START II also does not limit the number of nonstrategic warheads—active or otherwise—although these have been reduced through reciprocal unilateral initiatives. When nonstrategic and inactive nuclear warheads are included, this means that, even under START II, the United States and Russia will continue to possess some 10,000 total nuclear weapons each—even though only 3,500 can be deployed in strategic delivery vehicles.

These reductions and other arms control measures are important: the United States and Russia have unambiguously halted and reversed their bilateral nuclear competition. Yet as notable as these reductions and other measures are, they have not sufficiently altered the physical threat to either society. The reduced forces could still inflict catastrophic damage on the societies they target or could target. Much more can and should be done. Even if both START agreements are fully implemented, the level of nuclear forces will remain much higher than needed to meet the core deterrent function. Moreover, the START process has not yet addressed the many thousands of nonstrategic warheads and nondeployed strategic warheads that both sides would retain, which worsens the risks of breakout or theft and unauthorized use. The basic structure of plans for using nuclear weapons appears largely unchanged, with both sides apparently continuing to emphasize early and large counterforce strikes. And despite reductions in numbers and alert levels, both countries remain capable of rapidly bringing their nuclear forces to full readiness for use. This operational availability unnecessarily exacerbates the small but significant risk of erroneous or unauthorized use. Finally, some U.S. ballistic missile defense programs threaten to impede the arms control process by making uncertain the stability of deterrence at lower force levels.

Nuclear Force Levels and the Need for Deeper Reductions

After implementation of START II, the United States would have about 3,500 deployed strategic warheads, as well as a few hundred deployed nonstrategic warheads. In addition, the United States could hold as many as 5,000 weapons in reserve. Russia also would have about 3,500 deployed strategic warheads, assuming that it is able to provide the resources required to meet that level under the terms of the treaty. In addition, Russia is expected to retain several thousand nonstrategic warheads in its active stockpiles, plus an unknown number of such warheads in reserve. Thus, it can be assumed that, before the March 1997 Helsinki agreement to seek START III, each country planned to retain roughly 10,000 nuclear warheads into the early part of the next century. These reductions represent a substantial drop from the total of 60,000 to 80,000 U.S. and Soviet warheads at the peak of the Cold War. Implementation of a START III accord would continue these reductions to lower levels, but even further reductions are both possible and important to the security interests of both countries.

In addition to the fundamental reasons for such cuts discussed in Chapter 1, the Helsinki agreement to undertake additional reductions should improve prospects for Russian ratification of START II and for continued improvement in U.S.-Russian relations. START II benefits both U.S. and Russian national interests: it achieves rough numerical parity in deployed strategic warheads, and by eliminating the attractive targets presented by land-based intercontinental ballistic missiles (ICBMs) with multiple warheads, it achieves the important goal of enhancing stability by making a counterforce first-strike attack less attractive. From an American perspective, these security benefits alone should be sufficient to persuade Russia to ratify the treaty in its existing form.

START II, however, presents Russia with political and economic problems that have caused strong resistance in the Duma to ratification. The treaty requires Russia to destroy far more delivery vehicles than the United States. As a result, to maintain parity with the United States under START II, Russian defense officials assert that Russia would have to build and deploy more than 500 new single-warhead ICBMs at the same time as it is destroying hundreds of existing multiple-warhead missiles. But it makes no sense for Russia to spend scarce resources to build new nuclear weapons systems when smaller arsenals based on existing systems are more than adequate, even for existing nuclear missions.

Such a Russian nuclear buildup to meet START II levels would pose risks for the West by creating a serious domestic issue in Russia at a time of political uncertainty. In addition, even though these additional weapons would be permitted by the treaty, there could be calls in the United States to respond to what would be represented as a Russian program to field a new missile force containing the latest technologies. The framework agreement at Helsinki to seek reductions to a level of 2,000 to 2,500 deployed strategic warheads in START III is thus a welcome step.

Controls on Warheads

Although START II limits the number of warheads that can be mounted on strategic delivery vehicles, it does not limit the number or types of warheads that each side may possess. That is, under the terms of the treaty, each side can keep as many warheads as it desires—it is only limited in how many of those warheads may be mounted on long-range missiles or bombers, which are themselves limited by the treaty. It is therefore perfectly legal under START II to store and maintain for redeployment the warheads that must be removed from delivery vehicles to meet the treaty's limits. This failure to limit warheads, combined with the inherent capability of some delivery vehicles to carry many more warheads than START II permits, provides the possibility of rapid breakout. Russia or the United States could, for example, relatively quickly place additional warheads on land- and sea-based missiles and bombers. Indeed, Russian critics of START II have cited the breakout problem as a major reason for opposing its ratification.

Verifying limits on nuclear warheads is substantially more difficult than verifying limits on delivery vehicles. Current estimates of the total number of nuclear warheads in the Russian stockpile have a margin of possible error measured in the several thousands. This has been a key factor in discouraging efforts to establish warhead limits that are not associated with delivery systems. It is time, however, for the United States and Russia to step up to this challenge because deep reductions in nuclear weapons will be impossible unless the two sides lay a solid foundation today by agreeing to monitor warhead stockpiles and dismantling activities. It is encouraging that Russia and the United States agreed at the March 1997 Helsinki summit that initial efforts in this direction would be part of the START III agenda.

The very large numbers of nuclear warheads in today's nuclear arsenals leave little incentive to cheat, but as the numbers are reduced verification will become an increasingly important issue. Since nuclear weapons can be small and portable, and not easily detectable by technical means, a regime that would provide high confidence of locating a small number of hidden warheads would be extremely difficult to achieve. Even an imperfect verification regime would greatly reduce the uncertainties in present U.S. estimates of the number of Russian warheads, and enable reductions to proceed further than otherwise.

An improved verification regime would include a series of detailed data exchanges on the number and location of all nuclear warheads, fissile material production sites and their specifications, as well as total inventories of all fissile material stocks. These declarations would be subject to multiple random and challenge inspections of activities relating to storage and deployment, as well as to the dismantlement of warheads and the storage and disposition of recovered fissile material. An expanded inspection regime would profit by improvements in physical protection and accounting for fissile material.

In addition, as part of a general program of increased transparency, all historical records relating to production of fissile materials and weapons should be made available for review. Drawing on all of this new detailed information, classical intelligence techniques, including human sources, could also provide information with which to assess declarations and provide leads on possible diversions. While information from such classical sources, which might reveal even very small diversions, would probably prove to be extremely useful in a verification context and a powerful deterrent to cheating, its contribution cannot be assessed quantitatively.

With this amount of access, including challenge on-site inspections, the problem of detecting new production activities should be less demanding. In this connection, recent developments do provide an improved basis for unilateral remote detection of reprocessing or enrichment activities that would be required to produce new supplies of fissile materials.[4]

Although no single measure would guarantee against cheating, taken together such measures would create a web of access points and data, so that it would be increasingly difficult and risky to hide a strategically significant cache of weapons. Significant advances in recent years in the technologies for remote and proximate sensing, tamper-proof labeling, information processing, and long-range communication would in principle allow immediately verifiable monitoring of all declared nuclear warheads and fissionable materials—if the U.S. and Russian governments unreservedly agreed to collaborate in the comprehensive application of these technologies. Limited experiments in cooperative monitoring have already been undertaken under the Cooperative Threat Reduction Program (Nunn-Lugar Act) and serve to demonstrate its basic feasibility.

Even a comprehensive monitoring arrangement would not immediately resolve all uncertainty, since retrospective accounting for material would invoke a certain inherent degree of uncertainty. Given this uncertainty, it would be difficult to prove that no clandestine cache of weapons or materials had escaped the system unless evidence of inconsistencies or actual falsification in the data, or information on actual diversions through national classical intelligence sources emerged. Over time, however, the operation of a comprehensive monitoring arrangement can be expected to give either progressively improving confidence in its integrity or more specific reasons for doubt in the adequacy of the system in dealing with low levels of diversions or clandestine production. Without systematic cooperation and reciprocity, however, the effectiveness of monitoring will be severely limited.

Controls on Nonstrategic Warheads

Nonstrategic nuclear weapons generally have less sophisticated use controls and may be more vulnerable to theft and unauthorized use than strategic weapons, making an expanded program of reduction and control especially important for this class of weapons. START I and II address only strategic weapons; they contain no limits on the nonstrategic nuclear weapons beyond the ban on land-based missiles with ranges between 500 and 5,500 kilometers imposed by the Intermediate-Range Nuclear Forces Treaty. Even after the implementation of their reciprocal unilateral pledges not to deploy certain types of nonstrategic nuclear weapons and to dismantle some fraction of them, however, it is estimated that the United States will retain about 1,000 active and reserve nonstrategic warheads. The United States estimates that Russia will retain significantly more nonstrategic weapons. As noted in Chapter 1, Russia's abandonment of the Soviet Union's no-first-use pledge indicates a growing reliance on nuclear weapons, particularly nonstrategic weapons, at least in rhetorical terms.

The Nuclear Posture Review recommended that the United States retain nonstrategic nuclear weapons in order to reassure its NATO allies of the United States commitment to the defense of Europe. In the committee's opinion, however, U.S. nonstrategic nuclear weapons are not necessary for this purpose. The original rationale for their deployment—to deter or halt a Soviet invasion of Western Europe—has vanished. The NATO alliance, with its modern conventional forces and greatly increased strategic depth in Central and Eastern Europe, no longer needs to resort to threats of nuclear use to deter or repel invasion.

The related view that U.S. nuclear weapons must be retained in Europe to keep Germany from acquiring nuclear weapons also is of dubious validity. Germany was a leading proponent of the indefinite extension of the Nonproliferation Treaty (NPT) and has been outspoken in its support for additional controls on nuclear weapons and weapons materials. In the changed strategic context of the post-Cold War world, the strategic nuclear forces of the United States, and to a lesser extent of the United Kingdom and France, should provide abundant reassurance to their European allies against nuclear coercion without U.S. nonstrategic nuclear weapons stationed in Europe.

The committee suggested in Chapter 1 that the United States, in full cooperation with its NATO allies, should give serious consideration to seeking an agreement with Russia and other affected states that would prohibit the forward deployment of nuclear weapons in Central Europe. Foreclosing such deployments in a binding, reciprocal, and verifiable manner would be a clear signal that both Russia and NATO were committed to denuclearization of their relationship.

The situation in Asia in slightly different. As acknowledged when the weapons were withdrawn from South Korea in 1991 and removed from all naval vessels, U.S. military forces do not require nonstrategic nuclear weapons to deter attacks on U.S. allies in Asia or to defend those countries if attacked. North Korea, the only country in the region that is openly hostile to the United States, can no longer rely on Russian or Chinese support. Moreover, North Korea is covered by the most recent statement of U.S. negative security assurances to nonnuclear weapons states.[5] Senior U.S. officials continue to emphasize the potential destructiveness of a second Korean War, which places a particular premium on deterring any attack on South Korea, as well as on the maintenance of robust combined capabilities in the event deterrence fails.[6] The United States remains confident, however, that U.S. and South Korean conventional forces could defeat a North Korean attack, despite the destruction that such a war would entail.[7]

If the United States wishes to deter or respond to nuclear attacks on South Korea, Japan, or other Asian friends or allies, nonstrategic nuclear weapons would have no essential military advantages over strategic weapons, and the symbolic advantages historically attributed to nonstrategic weapons are outweighed by the political and security costs of the deployment of such warheads.

The presence of these weapons has provided allied countries with a basis for participating in nuclear planning and activity, which they regard as important to their security. Given this and the important symbolic role that forward-deployed nuclear weapons have played in the defense of U.S. allies throughout the world, the process of withdrawing nonstrategic nuclear weapons will require sustained high-level consultations.

Alert Levels

Despite the end of the Cold War, both the United States and Russia maintain the technical capability, even during peacetime, to launch thousands of nuclear warheads on short notice. This is particularly true of the United States, which maintains two-thirds of its submarines and virtually all of its land-based missiles in a high state of alert. Alert practices make deterrent forces immediately responsive, but they also increase the chances of an unauthorized or even accidental detonation of a nuclear weapon and contribute to the possibility of triggering a war that neither side intends. Current U.S. and Russian doctrines for nuclear operations also provide for extremely rapid response to evidence of attack, which risks having the decision to launch a nuclear strike made in response to an error in judgment.

The technical ability of each side to launch massive nuclear attacks with little warning creates incentives for the other side to prepare for virtually instantaneous decisions about a retaliatory strike before its nuclear forces or command system are destroyed. Russian military planners, for example, must still worry about the fact that accurate U.S. Trident and Minuteman (and, until START II is fully implemented, MX) missiles could destroy a large fraction of Russian nuclear forces and command-and-control systems with as little as 20 minutes warning of an attack, since only a small portion of the Russian submarine and mobile missiles are on patrol or positioned to survive a first strike. Especially troubling is that Russia, in protecting against the possibility of such a sudden attack, reportedly continues to rely on its capacity to launch ICBMs and pier-side submarine-launched ballistic missiles (SLBMs) on warning of a missile attack. According to some reports, the situation is exacerbated by the fragmentation and degradation of Russia's attack warning system. Thus, by deploying relatively large, lethal, and alert forces to deter the increasingly improbable circumstance of a deliberate surprise Russian attack, the United States may prompt Russia to adopt a posture that greatly increases the risk of erroneous or unauthorized launch. The strain on both countries would be relieved if neither had to worry about even the possibility of instant nuclear attack.

The issue is one of balancing risks. During the Cold War, reducing the risk of a surprise attack appeared to be more important than the risks generated by maintaining nuclear forces in a continuous state of alert. With the end of the Cold War, the opposite is now the more credible view, and this has important implications for U.S. nuclear policy, making dramatically reduced alert rates possible and highly desirable.

The United States and Russia have already taken some steps to reduce the alert status of their nuclear forces, but further action is needed. The challenge is to find ways not yet identified to further reduce or eliminate the capacity of nuclear forces to strike rapidly and with little warning without significantly decreasing their survivability or generating dangerous instabilities.

Targeting and Operational Doctrine

The 1994 U.S.-Russian detargeting initiative, which directed that the guidance systems of each country's missiles should no longer actively target the other, was a step in the right direction. But the missiles can be retargeted in a matter of minutes. With no indication to the contrary, the United States and Russia probably follow nuclear war planning procedures based on Cold War assumptions that emphasize the need for early massive strikes on nuclear forces and their command-and-control systems and attacks directed at national political and military leadership. This targeting doctrine was partially justified in the past under the general rubric of deterrence, which is said to depend on maintaining "forces of sufficient size and capability to hold at risk a broad range of assets valued by political and military leaders."[8]

Whatever may have been the rationale for this targeting doctrine during the

Cold War, its validity is dubious under current and near-term international security conditions in which the likelihood of all-out nuclear war is remote. Under these changed conditions, the arguments against attacking these targets take on much greater weight. Eliminating leaders and destroying communications would decrease opportunities to achieve early termination of hostilities; targeting policies calling for rapid destruction of nuclear forces and command centers create imperatives for the other side to launch vulnerable ICBMs and pier-side SLBMs before they are destroyed. As already noted, fear of such attacks encourages both sides to keep their nuclear forces at high levels of alert and could trigger a launch of nuclear forces in response to a false warning. A doctrine that provides for the rapid launch of nuclear forces cannot be justified in the foreseeable post-Cold War security environment, where the probability of an erroneous or unauthorized launch may be greater than the probability of a deliberate nuclear attack.

A policy of launching under attack poses unnecessary risks because it forces political and military leaders to make momentous decisions in a few minutes with incomplete information on the nature or origin of an attack. If both sides continue to maintain such options (and know that the other side does as well), this could increase the chance of miscalculation during a crisis.

The historical alternative to counterforce is countervalue targeting, in which the use of nuclear weapons is deterred by threatening to destroy concentrations of key industries, which are usually colocated with population centers, or by threatening to attack population centers outright. For its advocates the deterrent effect of countervalue targeting rests on the recognition that only a small number of nuclear weapons would destroy the cities, economy, and functioning society of even the largest country. For example, under certain conditions the detonation of as few as 20 nuclear weapons on Russian cities could kill 25 million people and destroy one-quarter of Russia's industrial output (see Box 2.2).

In addition to the moral objections to an explicit policy of targeting noncombatants, some have argued that countervalue targeting is not credible because destroying an opponent's cities would only result in destruction of one's own cities. A nation therefore might be deterred from retaliating against an opponent's cities in some circumstances unless its own cities had already been destroyed. Moreover, under such a policy the threat to use U.S. nuclear weapons in response to a nuclear attack on U.S. allies might not be considered credible. (Chapter 3 describes a policy of "adaptive targeting" that minimizes the pitfalls of both counterforce and countervalue targeting.)

Ballistic Missile Defenses

Somewhat paradoxically, the revolution in the offense-defense relationship wrought by the destructiveness of nuclear weapons discussed in Chapter 1 is likely to be a positive development because it negates the opportunity for national leaders to think of "winning" a nuclear war. If nuclear retaliation cannot be

BOX 2.2
The Destructive Power of a Few Nuclear Weapons

A simple calculation illustrates the destructive power of a small number of accurately delivered nuclear weapons. About 25 million people—one-sixth of the entire Russian population—live in 12 cities with populations greater than 1 million. These cities have a combined urban area of about 2,500 square kilometers. Detonation of a single 475-kiloton W-88 SLBM warhead from the U.S. arsenal would destroy an urban area of 100 to 150 square kilometers. Detonation of only 20 such warheads therefore could completely destroy the 12 largest Russian cities and kill 25 million people. Note that this calculation considers only the direct blast and thermal effects of the explosions; millions of additional deaths might result from firestorms, radiation (especially if weapons are detonated near ground level), and from disease and starvation resulting from the destruction of food and water supplies, hospitals, and other urban infrastructure.

Russia's industrial output is closely correlated with urban population but is somewhat more concentrated in the largest cities. Since 23 percent of Russia's urban population lives in the 12 largest cities, the attack postulated above would also lead to the direct destruction of at least one-quarter of Russia's industry. Indirect losses would be much larger.

prevented, either by defenses or by a truly disarming first strike, the rational employment of a large-scale nuclear attack for aggressive purposes is foreclosed. Missile defenses that offered a substantial capability to defend national territories would prompt other countries to increase and modify their offensive forces to compensate for the defenses. If compensating offensive deployments could be accomplished more quickly, cheaply, and reliably than the defensive deployments that prompted them, instead of providing a nationwide leak-proof shield, or anything close to it, the deployment of such defenses ultimately would result in the same threat of destruction but at a higher level of offensive forces.

The desire to avoid this situation was the rationale behind the 1972 Anti-Ballistic Missile (ABM) Treaty, which strictly limited the missile defenses allowed to the United States and the Soviet Union in order to prevent deployment of systems that would risk promoting arms races. In recognition of the continuing importance of this link between offense and defense, the Soviet side formally stated that U.S. withdrawal from, or material violation of, the ABM Treaty would be grounds for Soviet withdrawal from START I.[9]

The role that missile defenses can play for the United States is clarified by examining the kinds of missile threats facing it and its allies. A number of nations possess shorter-range (less than 500 kilometers) ballistic missiles for tactical purposes, such as the Scuds employed by Iraq during the Gulf War. Since the late 1950s, the United States has pursued missile defenses designed to intercept such short-range tactical missiles, which can be expected to be armed with non-nuclear warheads. Although technically challenging, this is a much easier task than intercepting longer-range missiles. The Clinton administration is focusing on upgraded Patriot (PAC-3) missiles and the Navy's sea-based Area Theater Defense system. These capabilities offer important benefits to U.S. military forces without threatening strategic stability. In addition, the United States has had considerable success in its efforts to control the proliferation of ballistic missiles and technology through the Missile Technology Control Regime, which now has most potential suppliers as members or adherents.

Advocates of expanded ballistic missile defense programs point to the possibility of longer-range theater ballistic missiles in the hands of potentially hostile Third World nations. Yet the potential threat posed to the United States and its interests by such missiles is limited to a few countries. North Korea has fielded modified Scud missiles that threaten all of South Korea, and it has tested and is continuing the development of a missile that could threaten Japan. Iran and Syria have purchased modified Scuds from North Korea that could reach Turkey, Saudi Arabia, and Israel; and Iraq might rebuild its missile program if international sanctions and monitoring were lifted.

All of these missiles have ranges of less than 1,000 kilometers. At present, the technologies required for longer-range missiles (multiple stages, powerful engines, special reentry vehicle materials, etc.) generally lie beyond the technical, although not necessarily financial, capability of these states to acquire and field them. A congressionally mandated independent panel, appointed by the Director of Central Intelligence to review a National Intelligence Estimate (NIE) and given full access to the relevant information, came to the following conclusion:

> . . . the Panel believes the Intelligence Community has a strong case that, for sound technical reasons, the United States is unlikely to face an indigenously developed and tested intercontinental ballistic missile threat from the Third World before 2010, even taking into account the acquisition of foreign hardware and technical assistance. That case is even stronger than presented in the NIE.[10]

The panel's report also states that the estimate "should have assured policymakers that this issue will receive continuing high priority, and that all possible technical alternatives will be investigated vigorously and that time to respond can be provided."[11] The committee notes that the NIE was criticized by the panel on a number of counts, including insufficient attention to other types of missile threats, such as cruise missiles and sea-based ballistic missiles. The criticisms did not cast doubt on the central conclusion about indigenously produced ICBMs but did raise questions about other potential missile threats to the United States.

Despite these uncertainties regarding the threat, in recent years the United States has begun conducting accelerated research and development on theater missile defenses, designed to intercept missiles with ranges up to 3,500 kilometers. The Clinton administration is developing two such systems for possible deployment—the Theater High-Altitude Advanced Defense (THAAD) system and the Navy's Theater-Wide Defense system. The United States is trying to persuade the Russian government that THAAD and the Theater-Wide Defense system can be deployed under a liberal interpretation of the ABM treaty, provided that they are not tested against ICBMs. The two governments tentatively agreed that a system with the characteristics of THAAD would not violate the treaty. But Russia demanded that any agreement must resolve all of the broader demarcation issues, including the line between more capable theater systems and strategic missile defenses. At the Helsinki summit, the two sides announced that they had reached agreement on all demarcation issues and instructed their negotiators to put the agreement in final form.[12]

Agreement within the U.S. government on the broader questions of strategic defense is still elusive, however. President Clinton vetoed legislation from the 104th Congress supporting immediate deployment of a nationwide defense against limited ballistic missile attack, but as this study was completed similar legislation was advancing in the 105th Congress.

If attempts to negotiate a mutually acceptable interpretation or revision of the ABM treaty were to fail and the United States unilaterally deployed systems perceived to be capable of intercepting Russia's missiles, Russia would probably take measures to ensure that its missiles could penetrate the defenses. Assuming that Russia had ratified START II, it could refuse to continue implementing the treaty; it would then be possible to maintain or deploy the multiple-warhead ICBMs that START II bans. Russia's ability to penetrate a U.S. defense could also be improved by increasing alert rates, which in combination with its reported plans to launch silo-based ICBMs and pier-side SLBMs upon warning of attack would increase the risk of an erroneous or unauthorized launch. It is also conceivable that Russia would deploy an ABM system in response that would force modification and expansion of the U.S. strategic forces. The collapse of START and the ABM treaty would undoubtedly mean a return to a more confrontational Russian military posture, making it more likely that such countermeasures would be taken—and needlessly increasing the risk of conflict between the United States and Russia.

The reactions of China, whose relatively small missile force is far more vulnerable to attack than that of Russia, also must be considered. China would undoubtedly see the prospect of even limited nationwide defenses as a threat to its deterrent, particularly if Russia responded to the U.S. initiative by deploying additional missile defenses. China, moreover, has stated that it would consider U.S. assistance to Japan in the deployment of a theater defense as providing Japan with a nationwide defense that would change the strategic balance in East Asia.[13]

China could respond with a significant expansion of its nuclear forces, which would increase the perceived threat to the United States, Russia, and India.

In short, deploying missile defenses outside the bounds of the ABM treaty could greatly diminish prospects for further reductions in nuclear weapons. The ABM treaty remains an essential part of the foundation of nuclear arms control. The United States and Russia will not continue to reduce their nuclear arsenals unless they are confident of the capabilities of their remaining deterrent forces. A possible strategic defense against a small number of nuclear warheads bought at the price of foreclosing further reductions in offensive nuclear arms—thus locking into place thousands of warheads capable of being aimed at the United States—would be a very poor investment.

THE OTHER NUCLEAR WEAPONS STATES: CHINA, FRANCE, AND THE UNITED KINGDOM

The nuclear forces of the three other declared nuclear weapons states, China, France, and the United Kingdom, have been in the range of 1 or 2 percent each of the weapons holdings of the United States or the Soviet Union/Russia (see Appendix B). Furthermore, with the end of the Cold War, British and French nuclear forces are making a transition to deploying almost all of their nuclear-armed ballistic missiles on nuclear submarines. China, with a much larger territory than the United Kingdom or France and with less tradition and capability in modern submarines, has maintained most of its nuclear warheads on land. Of these, only a few have intercontinental range; as for intermediate-range missiles, China possesses more than 60 land-based DF-3s, about 10 DF-21s, and one submarine with 12 missiles.[14] Presumably there are hundreds of bomber-delivered nuclear weapons.

NATO's two European nuclear powers, the United Kingdom and France, have already restructured their nuclear forces in response to the profound change in military threat. There is no reason to believe that rational defense planning and budget pressures would not keep these forces at the level of a few hundred warheads even if the United States and Russia chose not to achieve reductions from the START I levels. Similarly, Russian nuclear weapons that might be taken as threatening China have been very much reduced (and, in the case of the SS-20 intermediate-range missile, eliminated worldwide), and there is no theater nuclear threat other than by Russia that would seem to inspire China to expand its current nuclear capabilities.

The signing of the Comprehensive Test Ban Treaty (CTBT) by all nuclear states—including China, France, and the United Kingdom—is an indication that they have no intention of proceeding with qualitatively new systems. But over the decades China's announced policy has been transformed from a willingness to reduce its own forces when the United States and the former Soviet Union reduced theirs by perhaps half, to a promise that China would join in worldwide discussions after U.S. and Russian holdings are greatly reduced. Thus far the three other nuclear powers have not led in a movement to reduce the hazards posed by nuclear weapons, but achieving lower levels of nuclear forces will require their participation in the interests of stability and transparency.

NUCLEAR WEAPONS POLICY AND NONPROLIFERATION

Current U.S. nuclear nonproliferation policy can be loosely divided into three overlapping areas: (1) maintaining and strengthening the formal nonproliferation regime, (2) reassuring nations that foregoing nuclear weapons will not jeopardize their security, and (3) preparing to respond if additional proliferation occurred. The Clinton administration has shown leadership in consolidating the overwhelming support for the indefinite extension of the NPT and in achieving the CTBT. U.S. nuclear nonproliferation policy, however, could be made stronger in all three areas.

- The absence of change in U.S. nuclear posture and practice to reflect the dramatically altered post-Cold War conditions weakens the credibility of U.S. leadership on nonproliferation efforts.
- The uncertainty in U.S. positions regarding what assurances it will offer nonnuclear nations if they are threatened or attacked with nuclear weapons impedes efforts to constrain proliferation. The parallel U.S. failure to provide unambiguous assurances to nonnuclear states that they will not be subject to threats or employment of U.S. nuclear first use further weakens the nonproliferation regime.
- Counterproliferation policy, which mixes the problems of nuclear, chemical, biological, and missile proliferation and suggests additional roles for U.S. nuclear weapons in deterring so-called "weapons of mass destruction," enhances the potential appeal of nuclear weapons to others facing similar threats.

The Nonproliferation Regime

Proliferation of nuclear weapons was recognized as a major hazard to international security early in the nuclear age. The NPT, signed in 1968, sealed a complex bargain. It recognized the five countries that by that time had tested nuclear weapons (the United States, the Soviet Union, the United Kingdom, France, and China) as nuclear weapons states. All other countries signing the treaty would do so as nonnuclear weapons states and would agree not to develop or otherwise acquire nuclear weapons. The treaty forbade transfer of nuclear weapons technology and materials from the nuclear weapons states to the nonnuclear weapons states and enjoined the nonnuclear weapons states from accepting them. In return, the nuclear weapons states agreed not to transfer nuclear weapons to their nonnuclear weapons state allies and acknowledged the rights of the nonnuclear weapons states to enjoy the benefits of peaceful applications of nuclear energy under strict safeguards. The nuclear weapons states agreed to assist them in these peaceful activities provided that they did not undercut the underlying nonproliferation objectives of the treaty. The NPT also obligated the nuclear weapons states to proceed in good faith with efforts to reduce their nuclear arsenals, with the eventual goal of complete nuclear disarmament.[15] This provision was agreed to under strong pressure from some nonnuclear weapons states in order eventually to erase the treaty's discriminatory nature.

With its associated arrangements (such as the London Suppliers Group, which established voluntary nuclear export controls), the NPT has been a great success story. Forecasts of the 1950s and 1960s that predicted a world with dozens of nuclear powers before the end of the century proved wrong. These forecasts failed to appreciate that most countries would support nonproliferation because they recognize that their security is better served without nuclear weapons in the hands of their neighbors and possible adversaries. Some very difficult cases remain, but international sentiment in favor of the multilateral nonproliferation regime was again expressed in the spring of 1995 by the overwhelming support for indefinite extension of the NPT.

A major proliferation concern stems from the so-called "undeclared" nuclear powers—India, Israel, Pakistan, and (until recently) South Africa. These four states developed their limited nuclear capabilities during the Cold War but for reasons largely detached from the global East-West confrontation. The rationale for nuclear arsenals in each of the four undeclared nuclear powers was roughly the same: each perceived severe and persistent security threats on its immediate borders. Israel, India, and Pakistan have all refused to sign the NPT.

The worrisome problem of the undeclared nuclear powers notwithstanding, in recent years some notable instances of nuclear restraint and actual denuclearization have shown the effectiveness of nonproliferation efforts. New civilian governments in Brazil and Argentina ended years of preparations for nuclear weapons by agreeing in 1990-1991 to renounce such weapons and establish a bilateral inspection regime to verify this decision. In March 1993, South African President Frederik W. de Klerk admitted the existence of a nuclear weapons program but declared that the stockpile of six bombs had been dismantled before Pretoria joined the NPT in 1991. After extensive inspections with full cooperation by South Africa, the International Atomic Energy Agency (IAEA) subsequently accepted the declaration.

Of particular importance, after the breakup of the Soviet Union, Ukraine, Kazakstan, and Belarus joined START I and the NPT as nonnuclear weapons states, giving up forever their claims to the nuclear weapons of the old Soviet arsenal located on their territories. All of the nuclear warheads originally located in these newly independent states have now been returned to Russia. Had they retained these weapons, Ukraine and Kazakstan would have become the world's third- and fourth-largest nuclear weapons states respectively.

During the Cold War, South Korea, Taiwan, and Sweden, among others, considered developing nuclear weapons. That they did not do so is a testament to the effectiveness of a U.S. policy of judicious application of security assurances and guarantees. It is also the result, in no small part, of far-sighted regional leaders, bilateral nonproliferation efforts by the nuclear powers, and an increasingly robust international nonproliferation regime based on a growing worldwide aversion to nuclear weapons.

The international community has also demonstrated its strong commitment to nonproliferation in cases where signatories violate their NPT pledge not to develop nuclear weapons. When it was revealed after the Gulf War that Iraq had a massive clandestine program to produce nuclear weapons even while it was a member of the NPT and subject to full-scope IAEA safeguards, the United Nations Security Council approved a resolution that required intrusive measures to uncover, and authority to destroy, all vestiges of the Iraqi nuclear program, and instituted an indefinite monitoring program to provide continuing assurance that such activities had not restarted. The IAEA subsequently launched a program to improve the reach and efficiency of its safeguards, particularly in discovering undeclared activities and facilities.

The IAEA was soon tested when it uncovered evidence, supported by information from U.S. national technical intelligence, that North Korea had probably given a false declaration regarding its production of plutonium. For the first time in its history, the IAEA requested a special inspection of an undeclared site. North Korea responded by giving the required three-month notice that it would withdraw from the NPT. In the ensuing 18-month crisis, the United States managed a multinational diplomatic effort that culminated in a North Korean agreement to freeze its nuclear program and, over time, to dismantle its indigenous nuclear facilities in exchange for the acquisition of two safeguarded light-water reactors. Although this agreement was criticized in some quarters as bribery, it demonstrated, far better than sanctions or military action would have in this case, the commitment of the United States and the international community to prevent further nuclear proliferation.

Despite strong international support for the NPT, however, fundamental tensions remain embedded in the nonproliferation regime that could erode its effectiveness in the long run. Failure to continue progress on further nuclear reductions in support of Article VI could, over time, undermine the existing NPT consensus.

The Comprehensive Test Ban. The CTBT, which bans all nuclear weapons tests or other nuclear explosions, was formally opened for signature on September 26, 1996. Over the years the CTBT had become the primary litmus test of the sincerity of the nuclear weapons states' commitment to meet their Article VI obligations. As evidence of the strong international support for the treaty, by the end of 1996 a total of 138 countries, including the five nuclear weapons states, had signed it.

In addition to its political impact, the CTBT will serve as a barrier to further proliferation, albeit not a completely insurmountable one. Some types of fission weapons can be and have been developed successfully without tests, as demonstrated by the U.S. bomb dropped on Hiroshima (an entirely different design from the device tested by the United States in the world's first nuclear explosion in July 1945) and by the former South African weapons development program. The test ban, however, puts technical as well as political obstacles in the way of most states initiating and carrying out a program to develop simple fission weapons, and it will probably preclude development of boosted fission or thermonuclear weapons by threshold states. In addition, it places a high barrier in the way of development and production of new types of nuclear weapons by the nuclear weapons states.

The CTBT still faces a substantial hurdle before it is legally binding. In order for it to enter into force, all 44 countries with nuclear weapons programs or nuclear reactors must ratify it. India, which is one of the 44 states, has formally declared that it would never become a party to the CTBT because the treaty does not provide for the elimination of all nuclear weapons and because it makes India's signature mandatory. This effectively blocks the treaty's entry into force and the activation of its extensive verification regime.

Nuclear-Weapon-Free Zones (NWFZs). Efforts to ban all nuclear weapons from specific regions or environments have helped to build the nonproliferation regime and to limit the perceived utility of nuclear weapons. NWFZs constitute a useful complement to the NPT by further reducing the concern that potential adversaries in a zone might develop nuclear weapons, that nuclear weapons would become a symbol of national prestige in the region, or that nuclear weapons might be introduced into the region by outside nuclear weapons states.

The earliest successful NWFZ agreement was the 1959 Antarctic Treaty; other agreements put sea beds and outer space off limits to nuclear weapons. The first NWFZ in an inhabited area was the 1967 Treaty of Tlatelolco, which covers Latin America and the Caribbean. Tlatelolco was in large measure a response to the shock of the Cuban Missile Crisis, just as the 1986 Rarotonga Treaty, which covers the South Pacific, was spurred by anger over French nuclear testing in the region. In December 1995, 10 Southeast Asian heads of state voted to create a NWFZ to cover their countries. The most recent NWFZ agreement is the Pelindaba Treaty, which covers the continent of Africa. Opened for signature at a ceremony in Cairo in February 1996, the agreement builds on the lessons of Tlatelolco and Rarotonga, and proponents hope it will serve as a model for other agreements. When all the existing and new free-zone treaties take effect, nuclear weapons will be banned from all of the southern hemisphere except the open oceans and also from portions of the northern hemisphere.

Each of these agreements contains protocols intended to ban or limit the activities of the nuclear weapons states in the relevant region and to obtain commitments from them not to use or threaten to use nuclear weapons against the parties to the agreement. After a long delay, in 1981 the United States finally ratified Protocol II of Tlatelolco, which obligates all of the nuclear weapons states to apply the treaty to their territories in the region. In the case of the United States this includes Puerto Rico, Guantanamo Bay, and the Panama Canal Zone, where there are major U.S. military bases. In late 1995 the United States announced its willingness to support Rarotonga and signed the treaty's protocols jointly with the United Kingdom and France, although it has not yet ratified them. In 1996 the U.S. government formally agreed, "without any reservations," to the key protocols of the Pelindaba treaty governing the African nuclear-free zone, but these are also awaiting ratification.[16] Regarding Southeast Asia, some in the U.S. and other governments are reportedly concerned about the loss of naval transit of nuclear weapons across the zone, which was initially an issue for Tlatelolco as well.

Controlling Fissile Materials. The difficulty of acquiring fissile materials— highly enriched uranium (HEU) and plutonium—constitutes the principal technical barrier to the acquisition of a nuclear weapons capability. Fulfilling the goals of current arms control and keeping alive aspirations for much deeper reductions will depend on achieving much tighter controls on these materials.

Proposals for a global ban on the production of fissile materials for weapons have existed since the beginning of the nuclear age. Such a ban would serve both nuclear nonproliferation and arms control. A verified ban on production would cap or constrain the programs of the undeclared nuclear weapons states, which as non-NPT parties are not otherwise restrained, without requiring them to acknowledge or roll back those arsenals immediately. A cutoff would strengthen the nonproliferation regime by subjecting fissile material production facilities in all states to international inspection, thus removing another discriminatory aspect of the current regime. It would also help make the world safer for deep reductions by preventing any further legal accumulation of fissile materials for weapons purposes by the nuclear weapons states.

In addition, efforts to control fissile materials must address the problems presented by civilian use of fissile materials, particularly plutonium. In principle, virtually all mixtures of plutonium isotopes can be used to make nuclear explosives, and this committee's study of plutonium management concluded that the isotopic mixture produced by typical commercial power reactors is not much more difficult to use for bomb-making than is "weapons-grade" plutonium.[17] Thus, there is a tension between the rights implicit in Article IV of the NPT— which guarantees access to peaceful uses of nuclear energy to nonnuclear weapons states—and the underlying nonproliferation objectives set forth in Articles I and II, particularly as they relate to the dual-purpose technology of plutonium reprocessing. The NPT permits plutonium separation for nonnuclear weapons states, provided the separation facilities are subject to full-scope safeguards. A comparable dual-use issue applies to the rights to enrichment technology, since facilities that produce low-enriched uranium (LEU) for civilian nuclear power can be operated to produce HEU for research reactors—or nuclear weapons. These critical problems of dual-use technology demand increased attention to improving the safeguards and physical security for all civil plutonium and HEU.

Protecting Friends and Allies Who Forego Nuclear Weapons

The question of whether and how the United States and the other nuclear powers would provide for the security of countries who choose not to acquire nuclear weapons has been an issue since the 1950s. With the end of the Cold War and the demise of the bipolar international system, the question has acquired new prominence. The eagerness of Central and Eastern European states to join NATO reflects, in part, their anxiety about living alone in a nuclear-armed neighborhood. More broadly, some nonnuclear weapons states made the question of what assurances the five nuclear powers were prepared to provide an issue in the NPT extension conference. Positive and negative security assurances and guarantees (including no-first-use pledges) can work to decrease incentives for other countries to acquire nuclear weapons.

There are two basic types of security assurances and stronger security guarantees. Discussions of security assurances and guarantees sometimes confuse the two concepts. *Positive* assurances represent pledges from the nuclear weapons states that they would come to the aid of a nonnuclear weapons state if it were attacked or threatened with nuclear weapons. *Negative* assurances represent pledges that nuclear weapons states will not use or threaten to use nuclear weapons against a nonnuclear weapons state.

Positive security guarantees have usually meant that a nuclear weapons state would consider an attack on its ally, the recipient of the guarantee, to be an attack on its own territory, thus calling forth the use of its conventional and possibly nuclear forces in defense of the aggrieved party. From the U.S. standpoint, NATO is the premier example of treaty-obligated positive security guarantees. During the Cold War, the United States would have come to the defense of its NATO allies in Western Europe if the Soviet Union or any member of the Warsaw Pact had attacked them. The United States remains committed to such a response, including with nuclear weapons, if NATO members were attacked.

Japan and South Korea also enjoy positive security guarantees through bilateral treaties, which have played a major, some would say decisive, role in forestalling nuclear weapons proliferation in these countries. The United States has not given formal positive security guarantees to Israel, but the strength and endurance of the U.S. commitment suggests that that country is within the circle of full U.S. guarantees.

With the exception of the new entrants expected to join the NATO alliance over the next few years, it appears unlikely that the United States will extend new legally binding positive guarantees to other nonnuclear weapon countries in the foreseeable future. This does not mean, however, that the United States will be uninvolved in the security problems of the countries with which it develops significant cooperation, as it has with Israel. Although Ukraine sought but did not receive explicit guarantees from the United States in return for giving up its claims to the Soviet nuclear weapons left on its territory, the United States has pursued intense involvement in the development of Ukraine's relationship with the security system in Europe, through both bilateral defense cooperation and NATO's Partnership for Peace.

Negative security assurances have frequently taken the form of unilateral statements and other nonlegally binding instruments. There is a trend developing in the context of the NPT and the treaties on NWFZs to strengthen such assurances by recording them in the form of legally binding instruments. Current U.S. policy—first enunciated by the Carter administration in 1978 and most recently reiterated by President Clinton in April 1995 in connection with extension of the NPT—assures all nonnuclear weapons states that belong to the NPT that the United States will not use nuclear weapons against them except "in the case of an invasion or any attack on the United States, its territories, its armed forces or other troops, its allies, or on a State toward which it has a security commitment, carried out or sustained by such a non-nuclear-weapon State in association or alliance with a nuclear-weapon State."[18] The issue has been confused by the suggestion of some U.S. officials that nuclear weapons might be used to retaliate against the use of chemical and biological weapons, even if these are *not* perpetrated by a country "in association or alliance with a nuclear-weapon State."[19] Assuming the U.S. position is eventually not only clarified but transformed into an unambiguous posture of no-first-use of nuclear weapons, as the committee recommends in Chapter 3, the pattern of U.S. positive and negative security assurances in relation to nuclear attacks will include most of the countries of the world under its umbrella.

In the past, both China and the Soviet Union offered negative security assurances in the form of no-first-use pledges; that is, these nations declared that they would never use nuclear weapons first. As noted earlier, Russia has now backed away from that doctrine to something akin to the "first-use-if-necessary" policy that the United States and NATO maintained throughout the Cold War and still maintain today.

Counterproliferation: Preparing to Respond if Proliferation Occurs

From the beginning, U.S. policy to prevent the further spread of nuclear weapons has acknowledged that the United States should be prepared to deal with instances of proliferation wherever they might occur. Current U.S. counterproliferation policy includes efforts to achieve improved counterforce capabilities, active and passive defenses, improved intelligence, more effective export controls, support for various arms control agreements and the Cooperative Threat Reduction Program, enhanced counterterrorism capabilities, and changes in doctrine and declaratory policy to reflect new threats.[20] Certain aspects of the current policy, however, are likely to pose problems for efforts to constrain the role of nuclear weapons and to achieve greater arms reductions.

The first of these problems is that the policy reinforces an inappropriate linkage among the three categories of weapons—nuclear, chemical, and biological—commonly referred to as weapons of mass destruction. These weapons do share the capacity to provoke powerful fears and revulsion, which presumably have contributed to the constraints on their use since World War II. But nuclear weapons are in a class by themselves: they have an energy release roughly 1 million times that of conventional explosives for a given size and weight of munitions; they cause immense damage to the physical infrastructure of society as well as mass fatalities; their damage is immediate; their use leaves an unmistakable signature; and they have long-term radioactive effects, producing casualties at great distances over an extended period of time. As noted earlier, defenses against nuclear weapons are generally ineffective and may be counterproductive.

Weight-for-weight, chemical weapons (CW) are far less effective than nuclear weapons in causing fatalities and lack their immense physical destructiveness. Under certain circumstances, biological weapons (BW) might cause human casualties over a period of time comparable to those that would result from the use of nuclear weapons of equal size and weight, but in other situations might prove largely ineffective. In no event would BW produce the devastation of a target's physical infrastructure associated with nuclear weapons or, for that matter, with massive conventional attack. The effects of both CW and BW are much less predictable and much more subject to countermeasures than are the effects of nuclear weapons. For example, air filtration, as provided by gas masks, shelters, and vehicular systems, can provide protection against both CW and BW. Depending on the agent, various medical measures may be effective for prophylaxis and therapy.

Thus, chemical and biological weapons have limited value as weapons of war both because of their relatively unpredictable effects and because of the potential for defenses against them. If a terrorist group contemplated using chemical or biological agents again as a terror weapon, it is unlikely that nuclear weapons would be either a deterrent or a tool of choice in responding to such action.

The difficulties associated with designing, producing, and delivering the three types of weapons also vary substantially. Lumping them together blurs important distinctions that should guide policies to deal with the threats each poses and encourages both nuclear and CBW proliferation.

A second problem with counterproliferation policy has to do with the activities of aggressive states and terrorists. Some argue that, based on past behavior, one group of such states—Iran, Iraq, Libya, and North Korea—would be more willing to use these weapons and less affected by the traditional policies the United States has applied to deter such use by other countries. This leads to suggestions that preemptive actions may be necessary if these states are discovered to be close to acquiring a nuclear, chemical, or biological arsenal. Yet there is little historical evidence to suggest that the leaders of such countries are irrational in the sense of not having a cause-and-effect logic that shapes their decisions. One can be completely rational and still make catastrophic errors of judgment, as for example Saddam Hussein did by remaining in Kuwait in the apparent belief that the coalition forces would not take military action. Such leaders are thus not necessarily less susceptible to deterrence than others.

State-sponsored terrorism is a complicated case of the aggressive state problem. Any U.S. response would require confidence that one had the right sponsor, and calibrating that response to an appropriate level of force—especially if significant time passes before one is sure of the sponsor—could be extremely difficult. The use of CW, BW, and even nuclear weapons by truly independent terrorist groups is a genuine threat, as the Aum Shinrikyo chemical attacks in Matsumoto and the Tokyo subway illustrated. Such terrorism is often nihilist and indigenous, however, and the U.S. nuclear arsenal is largely irrelevant to combating it.

A third and final problem facing the architects of U.S. counterproliferation policy is the role of missile defenses, which have become a central element of current U.S. strategy. The issue has already been discussed in this chapter as it relates to U.S.-Russian relations. But it is the supposed threat posed by missiles in the hands of aggressive states—especially those with nuclear, CW, or BW capabilities—that has been the driving force behind much of the current U.S. interest in improved missile defenses. Missiles can have serious and destabilizing political and military effects and, as the Gulf War showed, powerful psychological impact. Some U.S. friends and allies, as well as U.S. forces overseas, could be vulnerable to missiles with ranges up to 1,000 kilometers, but they would also be at risk from other, more readily available means of delivery, such as aircraft, ships, or even land transport across borders.

Perhaps unintentionally, current U.S. counterproliferation policy suggests an almost complete reliance on U.S. unilateral action and exacerbates doubts that U.S. conventional military predominance will be sufficient to deal with threats posed by the proliferation of CW, BW, or nuclear weapons. Combined with hints that nuclear weapons might be used to respond to the use of CW or BW, this is a powerful message to weaker, more vulnerable nations about the apparent value of nuclear weapons—as well as the value of chemical and biological weapons. U.S. interests will be ill served by any policy that enhances the status of nuclear weapons, or that of CW and BW, and thereby increases incentives for their proliferation.

CONCLUSION

The United States has accomplished much to lay the foundations for stricter controls on and dramatic reductions in nuclear weapons, as well as fundamental changes in nuclear operations. But much more needs to be done by the United States and Russia, as well as by the other nuclear powers. The world looks to the United States, as the sole remaining superpower, for leadership. The agenda prescribed in succeeding chapters is ambitious and will not be accomplished quickly, but the time has come to intensify the effort to achieve it.

NOTES

1. Formally, the treaty limits each side to 6,000 equivalent warheads, a measure that substantially discounts strategic-bomber-delivered gravity bombs and air-launched cruise missiles.

2. The statement that the United States is not developing any new types of nuclear weapons has been challenged by critics who cite the B61-11 program. That program is officially designated as a modification, not a development. The "physics package" of the bomb remains unmodified for the B61, but the external envelope, including fusing and firing and various safety features, is being totally replaced as the bomb is reconfigured to become an earth penetrator.

3. In addition to producing plutonium, the present Russian reactors provide essential heat and electrical power for the neighboring communities. The United States and Russia are cooperating to convert the cores of these reactors to a fuel that produces significantly less plutonium and that will also be capable of safe storage.

4. Department of Energy, "Report of the Comprehensive Research and Development Review Committee for the U.S. Department of Energy Office of Nonproliferation and National Security," Washington, D.C., June 7, 1996.

5. See page 52 of this chapter for a discussion of current U.S. negative security assurances.

6. In congressional testimony in March 1996, the then-commander of U.S. forces in South Korea asserted that North Korea could strike Seoul "without moving a single piece of their vast forward arsenal" (General Gary E. Luck, to the Subcommittee on National Security, House Appropriations Committee, March 15, 1996). Authoritative U.S. policy documents further acknowledge that "a war would cause tremendous destruction on both sides of the DMZ [Demilitarized Zone], particularly in and around Seoul" (*United States Security Strategy for the East Asia-Pacific Region*, U.S. Department of Defense, Office of International Security Affairs, Washington, D.C., February 1995, p. 26).

7. Ibid., p. 26.

8. The White House, *A National Security Strategy of Engagement and Enlargement* (Washington, D.C.: U.S. Government Printing Office, 1994), p. 12.

9. "Statement by the Soviet Side at the U.S.-Soviet Negotiations on Nuclear and Space Arms Concerning the Interrelationship Between Reductions in Strategic Offensive Arms and Compliance with the Treaty Between the U.S. and USSR on the Limitation of Anti-Ballistic Missile Systems," June 13, 1991.

10. "Emerging Missile Threats to North America During the Next 15 Years," Independent Expert Panel Review of NIE 95-19. Unclassified version released by the Central Intelligence Agency to the Senate Select Committee on Intelligence, December 23, 1996.

11. Ibid.

12. The White House, Office of the Press Secretary, "Joint Statement Concerning the Anti-Ballistic Missile Treaty," March 21, 1997.

13. Since China has pledged never to use nuclear weapons against a nonnuclear state, this assertion may seem contradictory, but some Chinese analysts suggest that, given Japan's technical capability to build nuclear weapons if its long-standing policy against doing so changed, the acquisition of an

effective BMD system would give them a destabilizing combination of a shield and a potential nuclear sword.

14. International Institute for Strategic Studies, *The Military Balance 1996/97* (London: Oxford University Press for the International Institute for Strategic Studies, 1996), p. 179.

15. The text of Article VI reads as follows: "Each of the parties to the Treaty undertakes to pursue negotiations in good faith on effective measures relating to cessation of the nuclear arms race at an early date and to nuclear disarmament, and on a treaty on general and complete disarmament under strict and effective international control" (U.S. Arms Control and Disarmament Agency, *Arms Control and Disarmament Agreements: Texts and Histories of Negotiations*, U.S. ACDA, Washington, D.C., 1990, p.100).

16. At the same time, however, an administration spokesman said that Protocol I of the treaty "will not limit options available to the United States in response to an attack by an [African nuclear free zone] party using weapons of mass destruction," which suggests a major reservation to the non-use component of the treaty (White House, Text of Daily Press Briefing, April 11, 1996). The issue of using nuclear weapons to deter the use of chemical and biological weapons is discussed in detail in Chapter 3.

17. National Academy of Sciences, Committee on International Security and Arms Control, *Management and Disposition of Excess Weapons Plutonium* (Washington, D.C.: National Academy Press, 1994).

18. Warren Christopher, "Statement Regarding A Declaration by the President on Security Assurances for Non-Nuclear-Weapon States Parties to the Treaty on the Non-Proliferation of Nuclear Weapons," U.S. Department of State, Washington, D.C., Office of the Spokesman, April 5, 1995.

19. As legal justification for such a position some U.S. government officials cite the international law doctrine of "belligerent reprisal," which under certain circumstances justifies a "proportionate" military response that violates a treaty if the enemy has violated another treaty. Since first use of biological or chemical weapons is forbidden by the 1925 Geneva Protocol, a first use of chemical weapons by a country such as Libya might, under this argument, justify a response that violates another legal obligation—that is, the negative security assurances protocol of the African NWFZ. For a critical discussion of this issue, see George Bunn, "Expanding Nuclear Options: Is the U.S. Negating Its Non-Use Pledges?," *Arms Control Today*, May/June 1996, pp. 7-10.

20. William J. Perry, *Annual Report to the President and Congress* (Washington, D.C.: U.S. Department of Defense, March 1996), pp. 53-62. The 1993 Defense Counterproliferation Initiative was part of the effort to reorganize the Defense Department's forces and plans in the wake of the Cold War. To date, most of the department's effort has focused on increasing active and passive defenses for military forces subject to nuclear, chemical, or biological attacks.

3

A Regime of Progressive Constraints

The program recommended in this chapter would shift the focus of U.S. nuclear policy. Nuclear forces would be reduced; their roles would be more narrowly defined; and, while preserving the core function of deterring nuclear aggression, increased emphasis would be placed on achieving higher standards of operational safety. This shift would entail:

- further reductions in active weapons inventories;
- including all nuclear warheads in arms reductions;
- arrangements for exact, verified accounting and assured physical security of all warheads and fissionable materials;
- transforming the operational practices of active forces to eliminate continuous-alert procedures, commitments to rapid retaliation, and mass attack targeting; and
- reaffirming the integral relationship between restrictions on offensive and defensive systems.

As they make progress, the United States and Russia will want to engage China, France, and the United Kingdom on these issues and eventually make them full partners in the nuclear reductions process. And the regime of global nonproliferation agreements, including comprehensive new controls on fissile materials, will need to be cemented and expanded.

Taken together, these measures would transform nuclear force structure and operations as well as the ways that nations view the roles that nuclear weapons play in their national security policies. As agreed at the Helsinki summit, the United States and Russia should conclude an agreement to reduce to about 2,000 deployed strategic warheads each in a START (Strategic Arms Reduction Treaty)

III negotiation. Although important, this is only a first step. Efforts to begin transforming U.S. and Russian nuclear operations also should begin and need not await agreement on further force reductions. There are certainly links between reductions and changes in operations, but progress in one is not dependent on progress in the other. There should also be considerable flexibility in the transition from one stage of reductions to another, and the possibility of eliminating the dividing lines between stages should not be excluded. The committee has prescribed no time period in which each or all of the stages should be completed, since decisions on these matters depend on specific political and technological choices that cannot be foreseen now. The most important point is that the overall process should be structured to make it possible to proceed expeditiously to significantly lower levels of nuclear weapons, with dramatic changes in nuclear operations well in train.

A continuing high-priority effort is also needed to improve the protection of nuclear weapons and fissile materials in Russia. Joint U.S.-Russian work along these lines, which has been going on since 1991 under the Nunn-Lugar Cooperative Threat Reduction Program, complements and strengthens arms reductions and other changes in nuclear policies. (Because this committee and other NRC committees have recently offered detailed analysis and recommendations on this subject in other reports, the issue is not treated in detail here.)[1]

AN IMMEDIATE STEP:
TO 2,000 DEPLOYED STRATEGIC WARHEADS

At their March 1997 summit, Presidents Clinton and Yeltsin agreed that the next step after START II, with its level of 3,000 to 3,500 deployed strategic warheads, enters into force should be negotiation of a follow-on START III agreement reducing the number of deployed strategic warheads to 2,000 to 2,500 on each side. The committee believes that serious discussions of START III should begin immediately as part of the effort to encourage the Russian Duma's ratification of START II and also to reduce the Duma's current leverage over the arms control process. Formal negotiations were begun on START II and in that case led to a successful conclusion even though final ratification of START I had not been completed.

To move as quickly as possible to this reduced ceiling, the new agreements should operate within the existing technical frameworks of START I and START II. This will necessitate deferring for a brief time the introduction of certain key concepts that are critical to still deeper reductions, such as including all nondeployed and nonstrategic nuclear warheads in the overall verification and accountability of nuclear warheads. Reductions to the 2,000 level should be easily accommodated within the existing and anticipated strategic force structures of both sides without creating any operational or survivability problems and would more than adequately fulfill the core deterrent function for both sides.

President Yeltsin originally proposed a level of 2,000 at the beginning of the START II negotiations, but the United States concluded it was too large a step to take initially. As discussed in Chapter 2, a majority in the Russian Duma currently opposes ratification of START II. Now that Presidents Clinton and Yeltsin have agreed on a framework for a START III treaty, the impasse in the Duma could be resolved to the mutual advantage of both countries.

The committee expects that in START III both the United States and Russia will maintain ballistic missile submarines as the major component of a smaller number of nuclear delivery systems. Of the U.S. nuclear delivery systems that would remain under START II, ballistic missile submarines are the most survivable. They have proven to be dependable, are mobile, and can be concealed for long periods. They can roam the oceans of the world with little constraint and can remain at sea for an extended time. If the United States deployed half of a force of 2,000 warheads on submarines, and continued the current practice of having two-thirds of its submarines at sea, about 650 warheads would survive an attack, even if all bombers, land-based intercontinental ballistic missiles (ICBMs), and submarines in port were destroyed. Given the same reasoning, it can be assumed that Russia would continue to deploy a significant portion of its 2,000 warheads in a survivable mode.

In anticipation of the time when reductions to very low levels might take place, it would be useful at this stage to begin to involve the United Kingdom, France, and China. Their involvement, however, should probably be limited to informal exchanges in which the United States and Russia would keep the other nuclear powers informed of their progress. In turn, these states would have the opportunity to express their views. These exchanges will become more important when negotiations expand in the next step to include all strategic and nonstrategic warheads.

The remainder of this chapter offers prescriptions to resolve the questions raised in Chapter 2. These include recommendations (1) transforming further the U.S.-Russian relationship to increase operational safety and to make further nuclear arms reductions comprehensive; (2) strengthening the nonproliferation regime to provide reassurance to nonnuclear states and to respond to further proliferation should it occur; and (3) making much deeper reductions in the numbers of nuclear weapons.

FURTHER TRANSFORMATION OF
THE U.S.-RUSSIAN INTERACTION

Limiting All Nuclear Warheads

The need to shift arms control from its focus on delivery vehicles to include limits on warheads, acknowledging the verification challenge this raises, was noted previously. All nuclear warheads—regardless of type, function, stage of assembly, associated delivery vehicle, or basing mode—should be counted. Limits on the total inventory of nuclear warheads would minimize the reversibility of reductions and diminish the possibility of rapid breakout. Such limits would force the eventual dismantling of thousands of additional warheads, improve the stability of the nuclear balance, and demonstrate the commitment of the United States and Russia to very deep reductions.

Verifying limits on nondeployed and nonstrategic warheads would require transparency measures regarding the storage, production, and dismantling of nuclear warheads, as well as a mechanism for exchanging and verifying information about the location and status of warheads. These measures would go beyond those required to verify the limits on delivery vehicles and launchers in START I and II. In September 1994, Presidents Clinton and Yeltsin took the first step in this direction when they agreed in principle to exchange data on their nuclear arsenals and instructed their experts to meet to discuss what information could be provided to the other side. Efforts to implement these measures, and the transparency measures associated with the end of Russian weapons plutonium production were at an impasse as this study was coming to a close, so developing U.S.-Russian cooperation is still a formidable challenge. But it is apparent that once such cooperation is realized, the following information should be included:

- the current location, type, and status of all nuclear explosive devices and the history of every nuclear explosive device manufactured, including the dates of assembly and dismantling or destruction in explosive tests;
- a description of facilities at which nuclear explosives have been designed, assembled, tested, stored, deployed, maintained, and dismantled, and which produced or fabricated key weapon components and nuclear materials; and
- the relevant operating records of these facilities.

An exchange of this sort would be a valuable confidence-building measure even in the absence of a formal limit on warhead numbers, but the real value of a data exchange lies in its contribution to verifying such limits. Perhaps the simplest way to verify the data exchange would be to conduct both scheduled and unannounced inspections of nuclear weapons storage sites. Such inspections could begin with warheads slated for dismantling, move to warheads in the inactive reserve, and finally bring into the process weapons in the active stockpile (such as nonstrategic and bomber weapons in storage bunkers). Inspectors could verify the number of warheads at a declared site using relatively simple radiation detection equipment.

Even without a comprehensive and continuous warhead monitoring arrangement requiring full collaboration, inspectors with occasional access to declared facilities could verify the number of warheads present. Evolving technology will provide some improvement in the unilateral detection and surveillance of undeclared enrichment and reprocessing sites as well.[2] As a practical matter, all forms of agreed surveillance, from existing methods to the most advanced new technologies, would have to protect the secrecy of nuclear weapons design information, but this requirement can be met without undermining the legitimate purposes of transparency.

One particularly important focus of such an inspection process would be to verify that warheads removed from the declared stockpile for dismantling have indeed been destroyed and not simply moved to a hidden storage facility. In an earlier report, this committee described a straightforward verification scheme to achieve this goal, using perimeter-portal monitoring at dismantling facilities.[3] This scheme would count warheads as they entered the dismantling facility and would count "pits" (the basic nuclear component of warheads) as they exited, using intrinsic radiation or radiographic techniques. The pits would be stored under safeguards, initially bilateral, to ensure that they are not incorporated into new nuclear weapons, pending the ultimate disposition of the material.

Taken together, over time these measures could substantially increase U.S. and Russian confidence that remaining nuclear warheads and materials were accounted for. But no verification system could provide complete assurance that no clandestine stocks remained. Therefore, as nuclear reductions proceed to lower levels, the issue of how much uncertainty is acceptable becomes increasingly important. This, in turn, places a greater burden on the international security system to give confidence that there will be few incentives to cheat or that violations, when detected, will be dealt with swiftly.

Eliminating the Hair Trigger

In assessing the risks associated with nuclear arsenals, the operational and technical readiness of nuclear weapons for use is at least as important as the number of delivery vehicles or warheads. Elimination of continuous-alert practices should be pursued as a principal goal in parallel with, but not linked to, START III. It would reduce the perceived danger of short-warning-time attacks; it would make detecting preparations to use nuclear weapons easier and thereby increase the time available for political solutions; it would reduce pressures on command-and-control systems to stand ready to respond quickly and thus would decrease the chance of erroneous launch of nuclear weapons or a launch in response to a spurious or incorrectly interpreted indication of impending attack; it would allow both sides to increase barriers to unauthorized use of nuclear weapons; and it would enhance the political relationship by eliminating the assumption that the other side might launch a surprise attack.

Ideally, the launch readiness of nuclear forces would be reduced in ways that are readily transparent to the other side, so that both sides can be assured that a large-scale surprise attack is not possible. Care must be taken, however, to reduce launch readiness in ways that do not lead to instability. This requires that a portion of the force sufficient to satisfy the core function be able to survive any plausible attack. In addition, both sides must be convinced that neither could obtain a decisive advantage over the other by suddenly rushing to ready additional forces for use.

Reducing and eventually eliminating the possibility of surprise attack in a transparent and stabilizing fashion are a challenging but achievable goal. A small survivable force sufficient to satisfy the core function can be deployed on submarines at sea or mobile missiles out of garrison. Although this force need not—and should not—be operationally capable of rapid use, it might be difficult, at least in the near term, to demonstrate to the other side that these forces are incapable of prompt launch without compromising their survivability. However, the remainder of the force—silo-based missiles, mobile missiles in garrison, missiles on in-port submarines, and strategic bombers—can and should be rendered incapable of rapid launch in ways that would be readily verifiable. This has already been accomplished for bombers, by removing the nuclear bombs and air-launched cruise missiles and placing them in storage bunkers. In the case of ballistic missiles it is possible to remove warheads, shrouds, guidance systems, or other key components. Inspectors or remote monitoring devices could then verify that the systems had not been readied for launch and provide timely warning of any attempt to do so. A number of possibilities exist along these lines, and the defense establishments of the nuclear weapons states should be directed to develop a range of acceptable options as part of the reductions process.

Over the longer term, the United States and Russia, together with the other nuclear powers, should search for ways to assure each other that *all* nuclear weapons, including those on submarines at sea or on mobile missiles out of garrison, are incapable of being used quickly and without warning. The committee believes that it is possible to develop ways to do this while preserving stability and survivability. Although a number of means have been suggested for achieving this result (e.g., by having submarines patrol out of range of potential targets), this is an issue that requires detailed further study.

As a related confidence-building measure, the United States and Russia should adopt cooperative practices to assure each other that neither is preparing to launch a nuclear attack. Today, verification of alert status and warning of attack are provided solely by national technical means such as photo-reconnaissance, attack-warning satellites, and early-warning radars. All five nuclear weapons states could gain from an evolving program to share such intelligence with each other, or to install sensors (video cameras, seismic sensors, and the like) near the nuclear forces of other states to verify their status. A program to exchange military officers would also enhance confidence over time in the low alert rate and benign intentions of the other side.

Revising Targeting Policy and War Planning

The committee noted in the previous chapter that it makes little sense to preserve targeting plans that were developed to deter the Soviet Union, an adversary that no longer exists and whose successor state is in the process of dramatic change. The core deterrent function could be credibly maintained—and operational safety enhanced—while moving away from the concept of immediate overwhelming target destruction that dominated U.S. nuclear planning during the Cold War. In its aftermath, planning to retaliate massively against either military or civilian targets is not the appropriate basis for responsible decision making regarding the actual use or threat of use of nuclear weapons.

The United States should adopt a strategy that is based on much more selective target options and that would not require prompt attacks on counterforce targets or imperil major fractions of the nation's population either within or beyond the boundaries of the target area. Target planning might focus on major military facilities or core infrastructure such as energy network nodes located outside large urban areas, designed in all cases to minimize civilian casualties to limit the pressure for escalation and to allow political leaders to negotiate an end to nuclear attacks.

For decades U.S. nuclear war planning has focused on the articulation of a highly complex plan for nuclear war—the Single Integrated Operational Plan (SIOP). One original purpose of the SIOP—to integrate the target planning of the different U.S. armed services—remains valid. The SIOP contains thousands of targets and strictly prescribes U.S. nuclear operations through the various stages of a nuclear war. It incorporates many options, but its implementation is relatively inflexible. Now that the United States no longer faces a single, massive enemy with huge nuclear and conventional forces, the revolution in computation and communications makes possible a much more flexible strategy.

The committee concludes that U.S. national security would benefit from replacing the traditional SIOP concept with a much more flexible planning system of "adaptive targeting." Under this concept, U.S. military planners would retain and update lists of targets in potentially hostile nations with access to nuclear weapons. They would do so, however, under the presumption that nuclear weapons, if they were ever to be used, would be employed against targets that would be designated in response to immediate circumstances—and in the smallest numbers possible. Advance military planning and timely exercises are prudent and essential if national leaders are to have confidence in the dependability of their nuclear forces in a crisis. But there is a wide gulf between adaptive targeting and the present situation. Some changes in this direction have begun at the U.S. Strategic Command, but the move to an adaptive targeting approach should be accelerated and formally adopted.

Adaptive targeting would represent a natural complement to efforts to reduce the size and alert status of nuclear forces. The SIOP was constructed to coordinate a rapid attack by thousands of warheads against a well-defined enemy. As the size and alert status of nuclear forces change, and the probability of a massive Russian attack continues to fade, the United States will no longer require standing plans for a massive U.S. response. A dialogue between U.S.

and Russian military leaders on this subject would be useful and could help pave the way toward greater mutual understanding, which would facilitate deeper reductions in nuclear forces.

Relating Reductions and Ballistic Missile Defenses

A strong linkage exists between reductions in offensive forces and limits on defenses. This linkage was captured in the preamble to the Anti-Ballistic Missile (ABM) Treaty, in which the United States and the Soviet Union agreed that:

> Effective measures to limit anti-ballistic missile systems would be a substantial factor in curbing the race in strategic offensive arms and would lead to a decrease in the risk of outbreak of war involving nuclear weapons [and] . . . would contribute to the creation of more favorable conditions for further negotiations on limiting strategic arms."[4]

The committee has already noted that plans to develop and deploy systems intended to provide or capable of providing even limited national missile defense could weaken and possibly destroy the value of the ABM treaty. This would, in turn, threaten the deeper reductions in offensive nuclear arms that the committee recommends. The committee concludes, therefore, that the ABM treaty must remain "a cornerstone of strategic stability," as it was described by Presidents Clinton and Yeltsin at the conclusion of the Helsinki summit.[5] The ABM treaty is by no means a relic of Cold War thinking as some assert. On the contrary, it remains a logical adjunct of the continuing reality of offense dominance in conflicts involving nuclear weapons.

In a world in which the number of offensive nuclear arms is reduced drastically and the role of nuclear weapons is diminished, the ABM treaty will continue to play a crucial role. Opportunities to maintain and enhance its integrity will require periodic evaluation. Various technical constraints on tactical and national missile defense systems, always preserving the legitimate defensive capabilities against shorter-range missiles, can be consistent with the provisions of the treaty. For example, limits on the speed of interceptors or test warheads, intercept altitude, the number and geographical distribution of interceptors, sensor technology and integration, and the sale of technology to third parties should be investigated, and agreed interpretations should be negotiated in the Standing Consultative Commission.

Current U.S. counterproliferation policy puts great emphasis on the need for enhanced defenses against theater missiles. Some level of ballistic missile defense, in appropriate balance with other defensive measures, is desirable to defend U.S. forces and allies overseas from theater ballistic missiles. The focus of this activity should be to have available in the near future a mobile system capable of defending relatively small areas against projected theater ballistic missile threats, which the committee believes will be limited to the 1,000 kilometer range for some time.

NUCLEAR REDUCTIONS AND NONPROLIFERATION

This section examines several issues that involve interests and factors beyond the direct control of the United States and Russia. Some have been addressed in detail in the past but will need strengthening in an era of very small numbers of nuclear weapons. Others have been identified before but not thoroughly addressed; in any case they need to be revisited now in the new international circumstances. All of these issues can and should be addressed in ways that *both* enhance international security in the short term *and* support evolution toward a future world order in which security does not depend on the maintenance of national nuclear arsenals and explicit or implicit threats to use them.

Engaging the Undeclared Nuclear States

Nuclear weapons presumably held by the undeclared nuclear states—India, Israel, and Pakistan—pose a vexing problem. Their possession of nuclear weapons would become an even more troubling issue when the United States and Russia consider reductions to very low levels of warheads. Ways must be found to engage the undeclared states in a manner that would make it advantageous for them to move toward nuclear disarmament. Engagement in the process should not encourage, much less require, these states to declare their nuclear status, however, since this would likely be counterproductive. There are at least two major risks with open acknowledgment of the undeclared nuclear arsenals. The first is that such declarations could be destabilizing if there were to be a lag between the announcement and elimination. However widespread the belief in the undeclared nuclear capability might be, there could be political repercussions, such as calls for punishment or compensating measures by adversaries in the region, once the suspicions were publicly confirmed. The second is that open acknowledgment could run the risk of appearing to confer legitimacy or rewards, thus decreasing the nonproliferation benefit.

Reductions in U.S. and Russian nuclear arsenals and global nonproliferation initiatives, though helpful, will not suffice to engage the undeclared states. In the case of South Africa—the only country that has destroyed its entire nuclear arsenal—changes in the regional security environment (the withdrawal of Soviet-sponsored troops from neighboring states) and in domestic politics (the transition to majority rule) convinced the leaders of South Africa that its security was better served without nuclear weapons than with them. Achieving a similar result for the three remaining undeclared nuclear states will require a similar stabilization of their political, security, and perhaps economic situations. Over the long term, progress along these dimensions probably will be more important than the pursuit of initiatives related directly to constraining or eliminating the nuclear weapons programs of these countries.

Patient but persistent diplomatic strategies that are tailored to the security perceptions of each state will be required. Israel, for example, has already stated its willingness to enter into a nuclear weapon free zone (NWFZ) agreement provided that a comprehensive peace agreement for the Middle East is achieved.[6] The greatest contribution that the United States can make to promoting Israeli nuclear disarmament is to expedite the peace negotiations and to ensure that, at an appropriate point, these negotiations are linked with negotiations on an NWFZ. In the meantime, the United States should encourage full Israeli participation in global nonproliferation initiatives, such as the Comprehensive Test Ban Treaty and a fissile material production cutoff.

The cases of India and Pakistan are more complex in that their nuclear weapons programs are linked to each other and to that of China. It seems highly unlikely that India would agree to join the NPT while China's nuclear arsenal remains unconstrained by arms control. Long-standing Indian policy suggests that its ratification of the NPT would likely require a commitment by China to reduce and eventually eliminate its arsenal, as well as requiring additional improvements in Sino-Indian relations. The more difficult case is that of Pakistan, which has a history of armed conflict with India and whose conventional forces are numerically inferior to those of India. In addition to other initiatives to improve the regional security environment, prospects for South Asian nuclear disarmament could be enhanced by conventional arms control and confidence-building measures protective of Pakistani security.

This is a long-term challenge, and the United States should take the lead in attempting to stimulate the negotiations that might lead to more durable stability in South Asia. In addition, the United States should focus on nearer-term measures designed to reduce the chances for an expanded nuclear arms race or the use of nuclear weapons on the subcontinent. This would include regional agreements not to deploy, use, or threaten to use nuclear weapons or nuclear-capable ballistic missiles, together with continued efforts to engage India and Pakistan in global initiatives, including the CTBT and a fissile material production cutoff, as well as international controls on the civilian production and use of fissile materials.

Strengthening the Nonproliferation Regime

As noted in the previous chapter, Article VI of the NPT commits the signatories to work in good faith toward nuclear disarmament.[7] Achieving nuclear disarmament would require an international political order in which the possession of nuclear weapons would no longer be seen as legitimate or necessary to the preservation of national security, as discussed at greater length in Chapter 4. While building such an international order is very much a long-term project, a necessary even though not sufficient condition for its success will certainly be a continuing effort by the nuclear weapons states to reduce, systematically and progressively, the sizes of their nuclear arsenals and the roles that these play in their national security policies.

There are also important shorter-term links between arms reduction efforts by the nuclear weapons states and the prospects for nonproliferation. Perhaps most important, the short-term and medium-term effectiveness of the global non-proliferation regime requires the full support and cooperation of a large number of nonnuclear weapons states in the maintenance of a vigorous International Atomic Energy Agency with the inspection powers and resources required to do its job, the implementation of effective controls on the transfer of sensitive technologies, and the creation of transparency conditions conducive to building confidence that proliferation is not taking place. The degree of commitment of the nonnuclear weapons states to these crucial collective efforts can hardly fail to be affected by impressions about whether the weapons states are working seriously on the arms reduction part of the global nonproliferation bargain.

Some downplay the importance of nuclear weapons state arms reduction performance for nonproliferation by pointing to the many nonproliferation accomplishments that have been achieved *without* deep reductions by the nuclear weapons states, such as the termination of the nuclear-weapon programs of Argentina and Brazil, the relinquishing of nuclear weapons status by South Africa and three of the former Soviet republics, and the indefinite extension of the NPT in 1995. The committee believes this view does not give adequate consideration to the longer-term factors affecting this issues. Not only does it underrate the importance of the nuclear weapons states' performance for maintaining the active commitment to the nonproliferation regime of the large number of states that are not potential proliferants; it also fails to appreciate that none of the indicated nonproliferation victories is necessarily permanent, that the governments of many threshold states contain antibomb factions whose clout is strengthened or weakened by the actions of the nuclear weapons states, and that, most important, the world's expectations about what constitutes acceptable nuclear arms control performance by weapons states *after* the Cold War are likely to be different than they were when the Cold War was under way.

On this last point, while many members of the community of nations were probably not pleased with the immense nuclear arsenals accumulated by the United States and Russia during the Cold War, most understood that the characteristics of that deeply hostile and far-reaching confrontation constrained what could be expected from the two countries in the way of reductions in the sizes of those arsenals and the missions assigned to them. With the Cold War over, the world is likely to be impatient with U.S. and Russian maintenance of nuclear forces much more potent than the new circumstances seem to require. While the required majority for indefinite extension of the NPT was probably always assured, the essential consensus was only achieved by a combination of great diplomatic skill by the conference chair, the application of the immense political clout of the United States, and, crucially, a clear expectation of renewed commitment by the nuclear weapons states to faster, deeper, broader progress in nuclear arms limitations.

In addition to the reduction and dealerting steps discussed previously, three initiatives of the nuclear weapons states stand out as especially important in this regard: (1) achieving entry into force of the Comprehensive Test Ban; (2) extending NWFZs; and (3) expanding controls over fissile materials. All three would benefit the nonproliferation regime, as well as U.S. national security, in tangible as well as symbolic ways.

The Comprehensive Test Ban Treaty. Completing the text of the CTBT and opening it for signature represents a major nonproliferation achievement. Although the treaty cannot enter into force without the adherence of India, which is now adamantly opposed, the signatories (which include all of the nuclear weapons states) will be bound by customary international law not to violate the treaty's purpose. The overwhelming support the treaty received in the United Nations General Assembly in September 1996 and a growing number of signatories will create a powerful norm, which may well mean that there will never be another nuclear weapons test or other nuclear explosion. Moreover, with sufficient political will, the barrier against entry into force can be overcome, either by persuading India to sign or by relaxing the rigid entry-into-force requirement, which was included at the insistence of China, Russia, and the United Kingdom.

Controlling Proliferation Through NWFZs. Today, NWFZ agreements do not cover the regions with the greatest risks of nuclear threat, use, or proliferation—South Asia, Northeast Asia, and the Middle East. The effectiveness of such arrangements clearly depends on the participation of the states of greatest concern in the region; for example, Israel's participation would be essential for a successful agreement in the Middle East. But even partial agreements, as the Latin American NWFZ was for a long time, provide ways for states to take positive actions. In addition, they provide a regime in waiting for the day when conditions improve. The United States should support these agreements by signing the relevant treaty protocols promptly and without reservations.

A new NWFZ in Central Europe, perhaps including western states of the former Soviet Union, would offer immediate security advantages to Russia as well as NATO and states of the former Warsaw Pact. It could make accession to NATO by new states from the region more acceptable to Russia. At the same time, some former Warsaw Pact states and Soviet republics are seeking security assurances and guarantees that their nonnuclear status will not make them vulnerable to coercion or, in the worst case, aggression. A formal Central European NWFZ, coupled with negative security assurances from the nuclear weapons states, would help relieve these pressures and provide another basis for developing cooperative security arrangements in a region that for centuries has been subjected to innumerable invasions, occupations, and imposed territorial divisions. Achieving such an NWFZ in Central Europe will certainly require a reexamination of some aspects of collective security upon which the Atlantic Alliance is now based, as would the agreement to ban forward deployment of nuclear weapons in Central Europe discussed in earlier chapters.

Controlling Fissile Materials. Unprecedented international transparency and accountability for fissile materials are essential preconditions for achieving very low numbers of nuclear weapons. In the near- to medium-term, the United States can help lay the groundwork for these broader measures through two specific approaches.

The first is a worldwide ban on the production of fissile materials for nuclear explosives or the production of such materials outside international safeguards. A United Nations General Assembly resolution in 1993 called for negotiation of an international treaty of this kind at the Conference on Disarmament in Geneva. This would be the first such agreement that could include the undeclared nuclear powers, but so far these states have been reluctant to support it. The conference negotiations have been delayed because of resistance by the undeclared states to intrusive inspections and to freezing current stocks at unequal levels and because of the insistence of some nonnuclear weapons states that all states must acknowledge and account for their existing stocks. Although, at present, momentum for the start of serious negotiations has faded and early agreement is unlikely, a fissile material cutoff would be a significant nonproliferation measure and should continue to be strongly supported by the United States.

The second approach would address the problems presented by the civilian use of fissile materials, especially plutonium. U.S. leadership and active participation will be essential to improving International Atomic Energy Agency (IAEA) safeguards and to achieving measures for international civil plutonium management. Although highly-enriched uranium (HEU) is in some respects a greater proliferation risk, technical solutions for its management and disposition are straightforward and currently available.[8] In any case, international control of all civilian as well as military fissile materials will surely come to be seen as a necessary part of reductions to very low levels of nuclear warheads.

New agreements should extend the high level of security and accounting demanded for intact nuclear weapons—the "stored weapons standard"[9]—not only to all phases of the weapon disposition process but also to separated civilian plutonium and HEU worldwide. To this end, the United States should support transparency measures to declare all stocks of fissile materials worldwide, which would complement the declarations that all nonnuclear weapons states that are parties to the NPT are already required to provide to the IAEA.[10] The U.S. Department of Energy took an important step in this direction in February 1996 with the release of a comprehensive report on the production, import and export, and current stocks and location of U.S. plutonium and has urged Russia to do the same.[11]

In addition, current U.S. efforts to encourage the conversion of research reactors from HEU to LEU will relieve the dual-use problem by creating a demarcation between LEU as legitimate for research and power production and HEU as solely used for military purposes (weapons and naval nuclear reactors). Finally, although strong national and international export controls face problems of implementation and perceived discrimination, they do slow the spread of nuclear weapons materials and technology and are receiving increasingly strong international support through the voluntary London Suppliers Group and national legislation.

Reassuring States That Forego Nuclear Weapons: No-First-Use

The United States has not reassessed the array of positive and negative security assurances and guarantees it provided during the Cold War—and some it refused to provide—in order to bring these obligations in line with the dramatically changed international conditions. Any such reassessment will raise complex and difficult questions. How far is the United States willing to go in the defense of others—and how many others? How much flexibility is the United States willing to forego to build support for nonproliferation? Can U.S. conventional superiority be used to offer adequate deterrence and positive security assurances to replace the nuclear umbrella?

Taken together, positive and negative security assurances and guarantees help to reinforce the international consensus against the use of nuclear weapons. They also help reduce the incentives for other countries to acquire nuclear weapons. A commitment by the United States to maintain appropriately formulated positive and negative security assurances and guarantees, whether through defense cooperation or other means, is a fundamental element underlying the nonproliferation regime. Such commitments cannot be made lightly but, once made, will make a major contribution to stability.

It is probably not realistic at this time for the United States to be significantly more encompassing in its positive security assurances and guarantees, beyond those embodied in existing treaty commitments. Like the dilemmas faced by international collective security arrangements, clearly identifying who the aggressor is in a conflict and building the consensus to act against that nation may be difficult. The United States could, however, do more to make negative security assurances and guarantees serve its nonproliferation interests by constraining its own behavior—as negative assurances and guarantees do—in support of that cause.

To this end, the United States should announce that the only purpose of U.S. nuclear weapons is to deter nuclear attacks on the United States and its allies, adopting no first use for nuclear weapons as official declaratory policy. In the post-Cold War era, when nonproliferation is a high priority and the credibility of the nuclear powers' commitment to Article VI of the NPT is crucial to maintaining the international consensus behind the regime, a U.S. no-first-use pledge could help remove both reasons and excuses for proliferation. It would also assist with the dialogue with China and those nonaligned states that urged a no-first-use declaration during the negotiations on the NPT and the CTBT and now propose a no-first-use treaty.

A no-first-use declaration would recognize that, in the changed circumstances following the end of the Cold War, the United States should not threaten to use nuclear weapons to deter nonnuclear attack. Such a declaratory policy would be consonant with the committee's proposed restriction of U.S. nuclear weapons to the core function of deterring nuclear threats. It would not in any way suggest that the United States is less willing than in the past to come to the defense of treaty-bound allies in Europe or Asia.

U.S. positive security guarantees to such allies have been an important component of not only regional and international stability but also U.S. nonproliferation policy: they relieve such states, and by extension the neighbors of such states, of the need to consider developing independent nuclear arsenals. Changing to a no-first-use policy will require consultation with allies to reassure them that the United States will meet, by nonnuclear means, its obligations to come to their aid in the event of nonnuclear attack. The use of U.S. nuclear forces would be reserved solely for deterrence of and response to nuclear attacks. So long as the conventional military superiority of the United States and its allies remains largely unchallenged, the substantial benefits of a no-first-use policy would outweigh its small risks, provided the proper political groundwork is accomplished with NATO, South Korea, and Japan.

The recent change in Russian declaratory nuclear doctrine, from no first use to reserving a nuclear option in response to a conventional attack from any quarter, illustrates these transformed circumstances. The Russians now argue that they are at a serious conventional disadvantage vis-à-vis NATO and must therefore retain a nuclear first-use option.[12] They argue further that NATO membership for states in Central Europe exacerbates this problem, as it carries with it an obligation to permit the basing of U.S. or NATO nuclear weapons in these countries.[13] Such potential forward basing would feed Russian suspicions about the motives for NATO expansion; a negotiated ban on Russian and NATO forward basing of nuclear weapons, combined with the Central European NWFZ the committee recommends, would go far to allay these concerns, without reducing NATO security. A nuclear free zone could also help bring about a Russian recommitment to no first use, which is essential for achieving universal adherence to this standard.

The situation in Northeast Asia is similar. Despite the very real conventional threat posed by North Korea, the United States can achieve its deterrent and warfighting objectives on the Korean peninsula without recourse to nuclear weapons. The U.S. security guarantee to Japan is especially important in the nonproliferation context: despite the profound Japanese aversion to nuclear weapons, Japan clearly possesses the technical wherewithal to acquire them. A Japanese nuclear capability would seriously destabilize the Asian-Pacific region and deal a severe blow to the nonproliferation regime. But here again, given both the strength of the combined U.S.-Japanese and U.S.-South Korean conventional forces in the region, and the use by North Korea of U.S. nuclear threats as an excuse to acquire nuclear weapons, the threat of nuclear first use is both unnecessary and counter-productive for U.S. and allied security in the region.

Designing Responses to Future Proliferation

What actions should the United States take if one or more new states acquire nuclear weapons and use them to threaten or attack U.S. forces or allies overseas, or even the United States itself? To meet this possible future challenge, the United States needs to be prepared to take initiatives that will provide this country with greater leverage to impose sanctions on and otherwise coerce states that violate emerging norms of nonproliferation and nonaggression.

Responses Against Aggressive States and Terrorists. The current concept of "rogue" states emerged in U.S. policy circles in the late 1980s, driven by rising concern over the risks to U.S. interests from nations engaging in terrorism and aggression and seeking to arm themselves with nuclear, chemical, or biological weapons and the means to deliver them. More recently, in its support for the creation of the new Wassenaar Arrangement to control the diffusion of conventional weapons and dual-use technologies, the United States has made clear that Libya, Iran, Iraq, and North Korea are the primary focus of its concern.

Acquisition of nuclear weapons by any of these four states would be perceived to alter the political and military balance in its particular region. The impact would be moderated, however, because the same U.S. deterrence policy already in effect for other potential nuclear-armed adversaries would then apply to the newly nuclear states and their neighbors. Even without nuclear weapons, aggressive states could pose serious problems if they threaten U.S. friends and allies or if U.S. forces became engaged in a regional conflict. The United States should maintain armed forces adequate to meet its commitments and guarantee its own security with conventional arms.

Current U.S. policy tries to isolate those it considers aggressive states and, with varying degrees of success, attempts to persuade the international community to do the same. The continuing sanctions on Iraq in the wake of the Gulf War reflect an international consensus that Iraqi behavior is still unacceptable. U.S. efforts to persuade the international community that Iran deserves a similar isolation have not succeeded. In reality, most countries, including the United States, do not maintain consistently strict nonproliferation standards because nonproliferation concerns must compete with other bilateral or multilateral foreign policy interests. But U.S. interests would be best served by keeping up—and pressuring others to maintain—high standards in the handling of all nuclear technology exports to nonmembers of the NPT and to specific aggressive states.

Rather than opposing Russian assistance to the Iranian nuclear power program, for example, the United States should emphasize securing strict conditions for any cooperative and safeguarding activities. Like the North Korea case, sales of Russian equipment and technology represent a major opportunity to introduce international control of the fuel cycle in a country that might cross the line between civilian and military nuclear programs. The United States is supporting such conditions as the return of spent fuel to Russia, no-reprocessing and no-enrichment pledges, environmental monitoring, and formal agreement to any-time/anywhere IAEA inspections, which there is reason to believe that Iran would accept. By contrast, continuing U.S. attempts to prevent any nuclear trade with Iran are widely seen as contrary to Article IV of the NPT and as evidence that the United States believes that IAEA safeguards cannot be relied on to give timely warning of the diversion of fissile materials. It may also threaten wider U.S. relations with Russia, in particular gaining Russian cooperation on other important nonproliferation issues.

It is possible to construct scenarios for state-sponsored terrorists gaining access to a nuclear weapon. Detonation of a nuclear weapon in a U.S. city by terrorists would produce an immediate and overwhelming public demand for revenge, but nuclear response to nuclear terrorism where there is no established state sponsor is not feasible. And even for proven state-sponsored terrorists, the United States would have a range of options for retaliation, and a nuclear response should not be considered automatic for this case of nuclear use.

Responses Against Chemical and Biological Weapons Proliferation. One contentious area of current U.S. nuclear policy is whether nuclear weapons should be used to deter or respond to the use of chemical and biological weapons (CBW) by states against the United States, its military forces, or its allies. Some would have the United States enunciate an official policy of responding to CBW attacks with nuclear weapons, regardless of any negative security assurances to which it is committed. Others argue that the United States should make no explicit nuclear threat but allow or even encourage potential adversaries to assume the worst. This is the policy the United States followed in the Persian Gulf War. Former Secretary of State James Baker, for example, later wrote in his memoirs that at the time he "purposely left the impression that the use of chemical or biological agents by Iraq could invite tactical nuclear retaliation."[14] As already noted, subsequent U.S. statements made in connection with its signing of Protocol I of the African Nuclear Weapons Free Zone treaty and with Senate consideration of the Chemical Weapons Convention have maintained that ambiguity.

Yet neither ambiguity nor an outright policy of nuclear retaliation serves long-term U.S. goals or interests. As the committee argued earlier, the United States should state that it will use nuclear weapons only to deter and respond to the use of nuclear weapons by others. The United States does not need and should not want to employ nuclear deterrence to answer CBW threats.

A policy of nuclear deterrence of CBW would provide incentives and an easy justification for nuclear proliferation, which is inimical to U.S. security. Many other countries face far more plausible and immediate CBW threats than the United States. If U.S. policy points to nuclear weapons as the ultimate answer to CBW, other states could have an increased motivation to acquire nuclear arsenals. Highlighting new or continuing missions for nuclear forces could damage the nuclear nonproliferation consensus throughout the world.

The United States has other means to deal with the CBW challenge that do not have negative consequences for U.S. security. The most fundamental response is to be found in the Chemical and Biological Weapons Conventions, which outlaw both classes of weapons and have reinforced a taboo against their use that has held up remarkably well. Despite rhetoric to the contrary, this barrier to CBW proliferation and use remains strong and, with entry into force of the Chemical Weapons Convention and current moves to strengthen the Biological Weapons Convention, promises to grow stronger over time. In cases where states or nonstate entities ignore these conventions and threaten the use of CBW, the threatened states can often take reasonably effective measures to protect military or civilian personnel from the effects of CBW (in contrast to the case of nuclear threats). International pressure—United Nations resolutions or sanctions and other means—also can be brought to bear on states claimed to be producing, or about to use, such weapons.

U.S. conventional forces offer a formidable deterrent and war-fighting response to CBW. The threat of conventional retaliation against CBW use is far more credible than the threat of nuclear attack for other, even more compelling reasons. First and foremost, a policy of nuclear retaliation endorses the very methods the United States condemns: the use of weapons of mass destruction. It would likely invoke nearly universal condemnation, in fact, thus casting a U.S. adversary in the role of victim, whatever the act that provoked the United States. This would almost certainly be the case if the physical consequences of a nuclear response carried beyond the boundaries of the immediate target area or the borders of the opponent. Finally, it is difficult to imagine a provocation so extreme that any U.S. president would want to breach the threshold of nonuse of nuclear weapons, which after all survived even the extreme threats and tensions of the Cold War. Indeed, the worst outcome of a nuclear response is the prospect that it might be seen as militarily successful, thus inspiring renewed belief that the perceived efficacy of nuclear weapons warrants their retention or, worse, acquisition.

NUCLEAR FORCE REDUCTIONS: HOW LOW CAN WE GO?

Thus far this report has considered the broad operational and policy issues whose resolution should be part of the process of achieving the conditions for truly low levels of nuclear armaments. The final part of this chapter discusses two further stages of the nuclear arms reduction process beyond START III: first, the committee recommends a commitment to reduce nuclear weapons to a *total* inventory of about 1,000 warheads each; then the merits of still deeper cuts, to totals of a few hundred each in the U.S. and Russian arsenals, and the issues that would need to be addressed to make such deep cuts practical are examined.

Critics of proposals for deep nuclear arms reductions argue that such cuts could actually be counterproductive for a number of reasons:

- Large nuclear arsenals could help prevent proliferation. Potential pro-liferators will be discouraged from acquiring nuclear weapons because, as long as the declared nuclear weapons states, in particular Russia and the United States, maintain large arsenals, a potential proliferator could not possibly aspire to join the "big league" of the major nuclear powers. Once the nuclear weapons states agree to reduce their arsenals by substantial amounts, attaining relatively significant nuclear status will become easier, and hence potentially more attractive.
- Historically, possession of nuclear weapons has bestowed international prestige. By substantially reducing their nuclear arsenals, the United States and particularly Russia could find their relative prestige diminished.
- Since the major nations, including the United States, have renounced the use of chemical and biological weapons, nuclear weapons are the princi-pal deterrent against CW and BW threat and use. Decreasing nuclear weapons deployments amounts to unilateral disarmament vis-à-vis poten-tial BW and CW threats from rogue nations.

The committee believes that these arguments should not be a barrier to deep nuclear arms reductions:

- The motives of today's potential proliferators to acquire nuclear weapons are often determined by regional factors, as is discussed elsewhere in this report. The size alone of the nuclear weapons deployments of the major powers is unlikely often to be a significant factor in the decision of new states to seek a nuclear weapons capability.
- In the post-Cold War world, the committee believes the prestige of the United States is based on other factors that are more important than the size of its nuclear arsenal. Russia's longer-term status is far more depen-dent on its economic revival and political stability than on the size of its nuclear arsenal. With regard to nonnuclear weapons states, the important status attained by states such as Germany and Japan, and the lack of spe-cial status accorded to India, Israel, and Pakistan, as well as North Korea and Iran, countries that have or have sought nuclear weapons, should be noted.
- As discussed in Chapter 2, the committee believes that "weapons of mass destruction" is a misnomer that obscures militarily important differences among nuclear, chemical, and biological weapons. The issue of using

nuclear weapons to deter the use of CBW was discussed in a previous section, where the committee concluded that CBW threats can be deterred and effectively countered without relying on nuclear weapons.

Reducing U.S. and Russian Forces to 1,000 Total Warheads Each

Let us assume that the United States and Russia have achieved reductions to about 2,000 deployed strategic warheads each through a START III agreement. What should be the next step by the United States? We could move directly to the lowest possible level that would permit us to fulfill the core deterrent function. Or we could proceed more conservatively in steps to that goal. The committee chooses the latter for two reasons: first, in the present analysis the committee wants to provide the numbers—and the rationale for those numbers—at every stage in the reduction process to illustrate both its feasibility and its practicality; second, the committee proposes that, at the earliest possible time, the unit of account for nuclear weapons should become any nuclear warhead in the possession of a state, not just deployed strategic warheads. The problem of adopting such a unit of account is not the physical task of dismantling and destroying warheads but the far more difficult and crucial issue of accountability. Both the United States and Russia, and later other states, will need very high levels of assurance that all warheads have been included in such a regime and that remaining warheads can be accounted for. Between the 2,000 accountable strategic warheads of START III and the lowest level of deployed systems possible, the committee has chosen the level of about 1,000 total warheads in the inventory of each country as a logical intermediate stage.

Why only about 1,000 warheads? There are three important issues: (1) survivability; (2) the need to be able to perform the core deterrent function without question; and (3) the problem of the other declared nuclear powers.

Survivability. The general process of nuclear reductions outlined in this report involves a continuous effort to seek lower levels of risk and higher levels of U.S., allied, and global cooperation and security from nuclear attack. To achieve such benefits, however, the process must ensure stability at each rung of the ladder, and stability is most clearly guaranteed by the possession of survivable nuclear forces not at risk from a first strike. At a level of about 1,000 warheads, such survivability can be assured for the United States through the deployment of Trident ballistic missile submarines carrying appropriately downloaded missiles. This level also offers the Russian government the option to place a greater proportion of its nuclear forces in a survivable posture at sea or in land-mobile missiles out of garrison should it choose to do so.

Performance of the Core Function. The earlier discussion of targeting doctrine provides a second reason for seeking post-START III reductions with Russia to about 1,000 warheads. A force of this size would be able to maintain the core function satisfactorily against the most challenging potential U.S. adversaries under any credible circumstances, assuming that strategic defenses remain limited and transparent enough to avoid surprises. Nor does the committee see a need for a reserve nuclear weapons stockpile as a hedge against the emergence of new nuclear powers or clandestine expansion of the nuclear arsenals of existing nuclear weapons states.

The Other Declared Nuclear Powers. The nuclear arsenals of the other declared nuclear powers—China, France, and the United Kingdom—provide a third rationale for the proposed level of about 1,000 warheads. Given their current policies, discussed in Chapter 2, these countries would seem to pose no impediment to an otherwise desirable reduction of U.S. and Russian holdings to about 1,000 total warheads each. As long as these three countries pledge not to increase their nuclear forces and hold open the possibility of eventual reductions, the United States and Russia can reduce to a level of roughly 1,000 warheads without demanding reductions in their arsenals as a precondition.

Before examining the next stage in this reduction strategy, the committee wants to emphasize that its shift from accounting for delivery vehicles to warhead accountability does not suggest that delivery vehicles are no longer important. On the contrary, the accountability established in the INF treaty for missiles and in START I for missiles and bombers must be sustained and expanded. Verification measures for these systems have been tested and perfected over the years and should be continued.

Nonetheless, the accountability problems relating to a warhead count pose unique and significant difficulties. Information about some of the numbers and types of warheads maintained by the United States and Russia remains classified, although they have been described extensively in nongovernmental publications. The status and condition of warheads likewise relates directly to perceived requirements for a functioning system of nuclear deterrence. Even as the conditions set forth here for progress in this area are met, the committee admits that tough negotiations are ahead and even greater levels of U.S.-Russian transparency and cooperation will be required.

Other measures have already been undertaken to account for nuclear warheads and their components. In addition to the verification measures discussed in Chapter 2, a system of material protection, control, and accounting is being developed for components of the surplus weapons of the former Soviet Union as they are dismantled under START I. Funded by the United States under the Cooperative Threat Reduction Program established by the 1991 Nunn-Lugar legislation, this effort has the goal of reducing the risks of nuclear weapons proliferation, including such threats as theft, diversion, and unauthorized possession of nuclear materials. Some of the progress made and the systems developed could be directly applicable to a warhead accountability system for weapons remaining in military hands.

Reducing to a Few Hundred Warheads

Even the achievement of U.S.-Russian reductions to a mutually agreed level of about 1,000 total warheads should not represent the final level of nuclear arms. There will still be powerful reasons to continue down to a level of a few hundred nuclear warheads on each side, with the other three declared nuclear powers at lower levels or perhaps even with no remaining nuclear forces.

The case for reductions to a few hundred warheads each for Russia and the United States rests on the same basic arguments as that for reducing the numbers of nuclear arms in general. These deeper cuts will continue the process toward constraining nuclear weapons that was begun by earlier reductions. Reduction to below 1,000 total warheads each in the U.S. and Russian arsenals, as a major step toward mitigating the discrimination inherent in the current nonproliferation regime, will be significant for long-term success of the global nonproliferation enterprise. While the danger of unintended use of nuclear weapons would be smaller at a level of 1,000 warheads than it is now, moving to the level of a few hundred nuclear weapons would further reduce the risks of erroneous or unauthorized use. Finally, such reductions would further constrict the scope foreseen for any conceivable intentional use of nuclear weapons.

Reduction to 2,000 deployed strategic warheads could accomplished in a few years; while moving to 1,000 total warheads will take somewhat longer, it is still a bilateral action. To go on to a few hundred, however, will be a more complicated and multilateral process.

Conditions for Reductions to a Few Hundred Warheads. One particular measure the committee recommends as a precondition for low levels of nuclear forces should be emphasized: an even more effective warhead accountability regime. Verification of forces as low as a few hundred nuclear weapons requires a standard significantly more exacting than attainable by current capability and knowledge. While survivable nuclear forces at this level would offer each nuclear power important insurance against the covert retention or acquisition of illegal nuclear warheads by another state, the nuclear powers would certainly insist on reliable accounting of the residual existing warheads before they would agree to move toward such small arsenals. How well and how quickly the nuclear powers—especially Russia and the United States—are willing to account precisely to each other for the warheads they produced during the Cold War will go a long way toward determining the perceived feasibility of, and a realistic timetable for, reductions below 1,000 warheads.

This analysis does not assume a fundamental change in the nature of international relations in order to achieve low levels of nuclear arms. It does assume unprecedented cooperation and transparency among all classes of nuclear powers—first tier, second tier, and undeclared—on the specific issue of nuclear arms reductions. Success in engaging the undeclared states and the willingness of the second-tier nuclear powers to allow reductions—or even elimination—of their nuclear weapons will be essential. So will enhancements to the effectiveness of the nonproliferation regime. Such cooperation and enhancements, although conceivable with the present state of world politics, would become more feasible once the process of nuclear reductions and a growth of confidence in its basic soundness and stability, has proceeded for another decade or so.

The remaining nuclear forces would have to be survivable, their command-and-control structure would have to be redundant and robust, and widespread and effective national ballistic missile defenses must be absent. Moreover, even at this low level the committee does not see a need for a reserve nuclear weapons stockpile as a hedge against the emergence of new nuclear powers or clandestine expansion of the nuclear arsenals of existing nuclear weapons states.

The Size and Composition of Small Nuclear Forces. A level of roughly 300 warheads provides a somewhat arbitrary but nonetheless useful model for discussing this phase of arms reductions. From a purely technical point of view, roughly 300 nuclear weapons—of which at least 100 were secure, survivable, and deliverable—should be adequate to preserve the core function.

The committee recognizes that a progressive downward revision from the current levels of nuclear arms, and even from the lower numbers recommended in the previous stage, to a level of a few hundred deliverable warheads implies a drastic change in strategic target planning. A force of a few hundred can no longer hold at risk a wide spectrum of the assets of a large opponent, including its leadership, key bases, communication nodes, troop concentrations, and the variety of counterforce targets now included in the target lists. The reduced number of weapons would be sufficient to fulfill the core function, however, through its potential to destroy essential elements of the society or economy of any possible attacker.

Many suggestions have been made for force composition at the level of a few hundred, such as eliminating all intercontinental ballistic missiles, retaining only the strategic submarine force, or basing nuclear-capable aircraft in a dispersed mode. The committee does not recommend a particular approach, and it is likely and acceptable that different nuclear powers would choose different options, but one U.S. example is offered here to demonstrate the viability of the concept.

Consider a U.S. nuclear deterrent based only on submarines. Maintaining two survivable nuclear ballistic missile submarines (SSBNs) continuously at sea would require five in the inventory; at any given moment, two of the five may be assumed to be in port and one might be undergoing repairs or refitting. To maintain two fully survivable submarines at sea in both the Atlantic and the Pacific oceans would thus require 10 total SSBNs in the U.S. force. Assuming the current level of 24 missile tubes per Trident submarine, and a loading of one warhead per missile, this force would possess 240 operational warheads, of which some 100 would be kept at sea and survivable at any given time. Adding 60 additional warheads for spares, logistics, and refurbishment produces a total of 300 warheads in the inventory—one-third of which are survivable on a day-to-day basis.

This example assumes the use of current U.S. submarines and submarine operational procedures. In a period of heightened tensions that lasted up to several months, a larger fraction of the force could be deployed at sea, increasing the number of survivable weapons (and reducing the vulnerability of large numbers of nuclear weapons sitting on submarines in port). In the longer run, as the U.S. Trident submarines ended their operational lives, the United States could replace them with a generation of smaller submarines carrying fewer missiles, thereby increasing the number of survivable platforms held at sea.

In the operational posture of much smaller nuclear forces, the elements of the force would be designed for deliberate response rather than reaction in a matter of minutes. States could assure this result through transparency measures to make it clear that preemptive attack and instantaneous retaliation, including launch under real or perceived attack, are no longer feasible options. This might be achieved, for example, by separating weapons from delivery systems. Where this is impossible or extremely difficult, such as on submarines at sea, it can be achieved by limiting the range of missiles (by removing stages and/or adding ballast) and restricting the area in which the submarines might operate. In any case, the highest standards of protection against unauthorized use should be implemented on the nuclear delivery systems of all countries.

Other Issues at the Level of a Few Hundred Warheads. An infrastructure of nuclear weapons expertise sufficient to maintain the safety and reliability of the remaining nuclear weapons will be required. This infrastructure must be sufficiently transparent to provide accountability of the total number of nuclear weapons and to assure the international community that it is not being used for the development of additional types of weapons. Maintenance of such an infrastructure, including continued availability of highly capable technical personnel, should not be interpreted as contrary to achieving reductions. The reductions in nuclear forces advocated by the committee will also permit postponement of any decision on the preferred approach to meet further tritium requirements for several decades.

Managing and disposing of excess stocks of plutonium and HEU will pose a growing problem for the world. These excess fissionable materials derive from the dismantlement of surplus weapons stockpiles and from other military programs as well as from spent fuel in the civilian nuclear fuel cycle. While it will be a lengthy and continuing process, it is necessary and feasible to achieve the "spent fuel standard," in which all excess plutonium would be no more accessible and attractive for weapons fabrication than that in the spent fuel of commercial power reactors.[15] It will be necessary, of course, to ensure that commercial spent fuel is adequately protected worldwide. As discussed earlier, a related goal should eventually be to subject all stocks of fissile materials to the "stored weapons standard."

Strategic Stability at Low Levels. When examining the risk of small numbers of nuclear weapons, two issues must be addressed:

1. Can the transition down to a level of a few hundred nuclear weapons be made without crossing a zone of increased instability, which could increase the risk of preemptive attack in a crisis?
2. Will defenses enhance or decrease stability during the transition to small numbers of nuclear weapons?

The answer to the first question depends, most importantly, on the broader political and strategic situation prevailing among the nuclear weapons states. Reductions to the level of a few hundred nuclear weapons can be achieved without incurring instability by carefully managing the stages of the process, particularly the precise sequence in which delivery systems are reduced. In general, instabilities can be avoided if, among other measures, the more vulnerable systems are retired first, multiple-warhead missiles are eliminated (thereby decreasing the value of a single aim point), and the time at sea for submarine forces (which also must be controlled by effective permissive action links) is maximized.

Some will also ask whether ballistic missile defenses could eventually provide adequate insurance against deliberate or accidental launch of a small number of retained or clandestinely produced nuclear weapons once the nuclear powers move toward very small nuclear forces. Point defenses could have a limited value in increasing the survivability of any remaining fixed land-based nuclear strategic systems. But more elaborate defenses are incompatible with the program of major reductions, since such systems could at least be perceived as negating the deterrent value of a deployed force of a few hundred weapons. Moreover, given the large number of means available for delivering nuclear weapons, it is difficult to imagine a world in which clandestine delivery could be effectively prevented. Thus, it seems likely that the deployment of defenses capable of intercepting significant numbers of strategic ballistic missiles would prevent major arms reductions without adding to security.

Ballistic missiles designed for shorter ranges and the delivery of conventional munitions will probably remain in military arsenals while the process of nuclear arms reduction proposed here proceeds. Such ballistic missile deployments will in turn presumably motivate missile defenses designed to counter such regional and theater threats. The committee considers such deployments to be part of the continuing evolution of conventional military postures. The committee does not address the prospects for success of ongoing efforts to restrain the proliferation of ballistic missile technology, although it is important to recognize that states can rapidly convert conventionally armed ballistic missiles to carry nuclear warheads if such warheads are available. Thus, controls on nuclear weapons remain paramount. And any controls on ballistic missiles of a particular range must apply to all such missiles, not merely those their possessors claim are armed with nuclear warheads. For this reason, in the INF treaty the United States and the Soviet Union agreed to a worldwide ban on their possession of land-based missiles with ranges of 500 to 5,500 kilometers, whether nuclear armed or not.

How soon the United States and Russia could move to a level of a few hundred warheads, and the other three declared nuclear powers to equal or lower levels (or perhaps zero), would depend more on political than technical factors. Not only the United States but also Russia, for example, would have to perceive a growing momentum toward political and military stabilization in which cooperation with the other nuclear powers played a role. Russia's economic health, if not prosperity, would likewise be important to its willingness to proceed to a level that in nuclear terms would "equalize it" with the United States, United Kingdom, and France but also with China. The main technical issues regarding deep reductions are achieving an effective verification regime and maintaining survivability and the ability to reach targets as discussed above. By comparison with the political and other technical barriers—especially verification—to be overcome, the technical task of dismantling the additional numbers of nuclear weapons made surplus by such reductions is a modest one. Once the preconditions for a multilateral accord to reduce to a few hundred warheads were met, states could move relatively quickly from the 1,000 level to lower numbers of nuclear arms.

NOTES

1. For recent assessments of some of the major U.S. assistance programs in this area, see National Research Council, *Proliferation Concerns: Assessing U.S. Efforts to Help Contain Nuclear and Other Dangerous Materials and Technologies in the Former Soviet Union* (1997) and *An Assessment of the International Science and Technology Center* (1996). Two studies by the National Academy of Sciences' Committee on International Security and Arms Control of the particular problem of excess weapons plutonium are *Management and Disposition of Excess Weapons Plutonium* (1994) and *Management and Disposition of Excess Weapons Plutonium: Reactor-Related Options* (1995). All were published by the National Academy Press, Washington, D.C.

2. Department of Energy, "Report of the Comprehensive Research and Development Review Committee for the U.S. Department of Energy Office of Nonproliferation and National Security," Washington, D.C., June 7, 1996.

3. "Under such an arrangement the disassembly facility would be securely fenced, with the exception of monitored entry and exit points. At the entry point, technical equipment could be used to verify that an entering object is a nuclear weapon. A variety of technical means to do so exist that could be used in a mutually supportive manner. . . . At the exit point, the material going out could be assayed for fissile material content (by methods external to the canisters containing the fissile components, to avoid inspection of the detailed dimensions of the components, which itself is classified information)" (*Management and Disposition of Excess Weapons Plutonium*, op. cit., p. 106).

4. U.S. Arms Control and Disarmament Agency, *Arms Control and Disarmament Agreements: Texts and Histories of Negotiations* (Washington, D.C.: U.S. ACDA, 1990), p. 157.

5. The White House, Office of the Press Secretary, "Joint Statement Concerning the Anti-Ballistic Missile Treaty," March 21, 1997.

6. Israel introduced a draft resolution on October 31, 1980 to the United Nations General Assembly proposing the establishment of a NWFZ in the Middle East. The resolutions calls for convening a conference of all Middle East and adjacent states with a view to negotiating a NWFZ. In later exchanges on the subject Israel made it clear that successful completion of the "Peace Process" was a precondition for accepting the NWFZ. Israel has also indicated that a regional approach to nuclear arms control is to be preferred over accession to the NPT. For an extensive discussion of the nuclear weapons status and prospects in the Middle East, see Shai Feldman, *Nuclear Weapons and Arms Control* (Cambridge, Mass.: MIT Press for the Center for Science and International Affairs, Harvard University, 1997).

7. *Arms Control and Disarmament Agreements*, op. cit., p. 100.

8. HEU is of particular concern because it can be used in a simple gun-assembly-type nuclear device and hence would be attractive to a wider range of would-be proliferators. But HEU can be diluted with other naturally occurring isotopes of uranium to make low-enriched uranium (LEU), which cannot sustain the fast-neutron chain reaction needed for a nuclear explosion. LEU is the fuel used for most of the world's nuclear power reactors. The technology required to return LEU to weapons-grade uranium is costly, time consuming, and not readily available.

9. *Management and Disposition of Excess Weapons Plutonium*, op. cit., p. 31.

10. Note that India, Israel, and Pakistan are not members of the NPT, which means they are exempt from the treaty's requirement for full-scope safeguards on their nuclear programs.

11. U.S. Department of Energy, *Plutonium: The First 50 Years* (Washington, D.C.: U.S. DOE, February 1996). Unlike the United States, Russia also has stocks of separated reactor-grade plutonium, which would need to be included in any such declarations.

12. "Strategic Forces Now in the Forefront of Russia's Defence— Commander," BBC Summary of World Broadcasts, Part I, the Former USSR, December 19, 1996, No. SU/2799, pp. S1/1-S1/2.

13. NATO issued a statement on December 10, 1996 that it had "no intention, no plan, and no reason to deploy nuclear weapons on the territory of new members" ("Final Communique Issued at the Ministerial Meeting of the North Atlantic Council," NATO Press Communique M-NAC-2 (96)165, December 10, 1996).

14. James A. Baker III with Thomas M. Defrank, *The Politics of Diplomacy* (New York: GP Putnam's Sons, 1995), p. 359. This ambiguity was conveyed to Saddam Hussein in a letter from President Bush just before the start of the Gulf War. The relevant passage reads: "The United States will not tolerate the use of chemical or biological weapons, support of any kind for terrorist actions, or the destruction of Kuwait's oilfields and installations. The American people would demand the strongest possible response." ("Confrontation in the Gulf: Text of Letter from Bush to Hussein," *The New York Times*, January 13, 1991). Despite the U.S. warning, Iraq did undertake destruction of the Kuwaiti oilfields.

15. *Management and Disposition of Excess Weapons Plutonium*, op. cit., p. 34.

4

Prohibition of Nuclear Weapons

The end of the Cold War has created conditions that open the way for consideration of proposals to prohibit the possession of nuclear weapons. The committee recognizes that it is not clear how or when this could be accomplished. Fundamental changes in international politics would be a precondition for comprehensive nuclear disarmament, and this is not something that can be forced into an arbitrary timetable. Nonetheless, for the same reasons that the committee recommends rapid and substantial reductions in the size, readiness, and salience of national nuclear arsenals, the time also has come to begin to devote serious attention to the prospects for prohibiting those arsenals and to fostering the conditions that would have to be met to render prohibition desirable and feasible. Although the reductions recommended in Chapter 3 are logical steps on the path toward comprehensive nuclear disarmament, the final step of banning nuclear weapons should only be undertaken in circumstances such that, on balance, it would enhance the security of the United States and the rest of the world.

The committee uses the word "prohibit" rather than "eliminate" or "abolish" because the world can never truly be free from the potential reappearance of nuclear weapons and their effects on international politics. Even the most effective verification system that could be envisioned would not produce complete confidence that a small number of nuclear weapons had not been hidden or fabricated in secret. More fundamentally, the knowledge of how to build nuclear weapons cannot be erased from the human mind, and the capacity of states to build such weapons cannot be eliminated. Even if every nuclear warhead were destroyed, the current nuclear weapons states, and a growing number of other technologically advanced states, would be able to build new weapons within a few months or few years of a national decision to do so.[1] A regime for comprehensive nuclear disarmament must, therefore, be embedded in an international security system that would make the possibility of cheating or breakout highly unlikely.

THE BENEFITS AND RISKS OF NUCLEAR DISARMAMENT

In exploring the desirability and feasibility of prohibiting nuclear weapons, the balance of benefits and risks that this course of action would entail must be evaluated. A durable prohibition on nuclear weapons would have three main advantages. First, it would virtually eliminate the risk that nuclear weapons might be used by those states now possessing them. Even the smallest nuclear arsenals have immense destructive power. No matter how carefully and conscientiously these arsenals are constructed and operated, there will be some risk that they might be used, either deliberately or accidentally, authorized or unauthorized. In a sense, a prohibition on the possession of nuclear weapons is a logical extension of the dealerting measures recommended in Chapter 3, extending from hours or days to months or years the time required to reconstitute an ability to use nuclear weapons. A durable prohibition would expand as far as possible the firebreak between a decision to ready weapons for use and the ability to launch a nuclear attack, thereby allowing as much time as possible to resolve the underlying concerns, and decreasing the risk of nuclear catastrophe to an irreducible minimum.

Second, a prohibition on nuclear weapons would reduce the likelihood that additional states will acquire nuclear weapons. Although the Nonproliferation Treaty (NPT) currently enjoys almost universal adherence, the nuclear weapons states cannot be confident of maintaining indefinitely a regime in which they proclaim nuclear weapons essential to their security while denying all others the right to possess them. The recognition that a permanent division between nuclear "haves" and "have-nots" is unacceptable is captured in Article VI of the NPT, in which all parties promise to pursue complete nuclear disarmament. This commitment was reaffirmed by the United Nations Security Council in connection with the 1995 NPT extension conference. In a recent advisory opinion, the International Court of Justice (ICJ) underscored the "vital importance" of satisfying this obligation under Article VI:

> In the long run, international law, and with it the stability of the international order which it is intended to govern, are bound to suffer from the continuing difference of views with regard to the legal status of weapons as deadly as nuclear weapons. It is consequently important to put an end to this state of affairs: the long-promised complete nuclear disarmament appears to be the most appropriate means of achieving that result.[2]

Moreover, the current lack of a serious commitment to comprehensive nuclear disarmament undermines the authority of the United States and other nuclear weapons states in combating proliferation and responding to violations of the NPT. It would be easier to marshal decisive international action against countries attempting to acquire nuclear weapons if a global prohibition on the possession of such weapons were in effect.

A third advantage of comprehensive nuclear disarmament has to do with the uncertain moral and legal status of nuclear weapons. In the advisory opinion cited above, the ICJ unanimously agreed that the threat or use of nuclear weapons is strictly limited by generally accepted laws and humanitarian principles that restrict the use of force.[3] Accordingly, any threat or use of nuclear weapons must be limited to, and necessary for, self defense; it must not be directed at civilians, and be capable of distinguishing between civilian and military targets; and it must not cause unnecessary suffering to combatants, or harm greater than that unavoidable to achieve legitimate military objectives. In the committee's view, the inherent destructiveness of nuclear weapons, combined with the unavoidable risk that even the most restricted use of such weapons would escalate to broader attacks, makes it extremely unlikely that any contemplated threat or use of nuclear weapons would meet these criteria.

Nuclear disarmament poses risks as well as benefits, however. First, there is the risk that the prohibition on nuclear weapons might break down. States might cheat if they believed that small nuclear arsenals could be used successfully for coercive purposes. States might also be impelled to withdraw from a comprehensive nuclear disarmament agreement if, at some point, they believed their vital interests could no longer be protected without nuclear weapons. To reduce these risks, a disarmament regime would have to be built within a larger international security system that would be capable not only of deterring or punishing the acquisition or use of nuclear weapons but also of responding to aggression of all kinds. This system would have to be structured so that no nation could believe that either it or any other state could obtain significant and long-lasting advantages from building or brandishing nuclear weapons or from nonnuclear aggression for which only a nuclear capability would serve as a deterrent. In a subsequent section examples are given of the sorts of arrangements that might be useful and necessary to meet this challenge.

Second, there is the concern that comprehensive nuclear disarmament would remove the moderating effect that nuclear weapons have had on the behavior of states, resulting in an increased risk of major war. The nuclear era represents the longest period without war between the major powers since the emergence of the modern nation state in the sixteenth century.[4] More than 100 regional conflicts, including civil wars, have been fought since the beginning of the nuclear age, but none of these conflicts generated direct combat between the nuclear weapons states. It is reasonable to assume that the cautionary effect of nuclear weapons is at least partially responsible for this absence of major wars. Thus, it is argued, if the major powers believed that the risk of nuclear war had been eliminated, they might initiate or intensify conflicts that might otherwise have been avoided or limited. Complete nuclear disarmament might lead, then, to the frequent, large-scale conventional conflicts that characterized the prenuclear era, with the additional risk that one or both sides would acquire and use nuclear weapons during a protracted war.

There is, however, no demonstrable relationship between the actual possession of nuclear weapons and the avoidance of war. First, even if all nuclear weapons were eliminated, the inherent capacity of major powers to build nuclear weapons would act as a deterrent to the outbreak of major conventional wars, since both sides would fear that the other might acquire and use nuclear weapons during a protracted struggle if its vital interests were threatened. In other words, existential nuclear deterrence, as discussed in Chapter 1, would remain to some extent even if nuclear arsenals were dismantled. Second, there have been, and continue to be, profound changes in the structure of the international order that reduce the probability of major war, independent of nuclear deterrence. These include the spread of democracy; the growth of information-based economic systems that do not depend on or benefit from territorial conquest; expanding economic interdependence and integration; the emergence of strong international political and financial institutions, such as the United Nations and the International Monetary Fund; the diffusion of global communications and shared culture, which limit the degree to which governments can control information and propagate negative images of adversaries; the advent of modern intelligence and surveillance systems that facilitate accurate assessments of military capabilities and which make surprise attacks less likely to succeed; the development of collective security arrangements, such as NATO and the Organization for Security and Cooperation in Europe; and, more recently, deployment by the Western powers of modern conventional armaments, such as precision-guided munitions, which improve the effectiveness of defenses against armored attacks. In short, the avoidance of major war in the nuclear age can be attributed to many factors rather than to nuclear deterrence alone. It is not unreasonable to believe that a continuation of the trends mentioned above, together with the development of more robust collective security arrangements, the maintenance of modern and capable conventional forces, and the deterrence provided by the capacity of major states to build nuclear weapons, could be capable of deterring large-scale war among the major industrial powers just as effectively as the current system—and with fewer risks.

After considering these risks and benefits, the committee has concluded that an essential long-term goal of U.S. policy should be the creation of international conditions in which the possession of nuclear weapons would no longer be perceived as necessary or legitimate for the preservation of national security and international stability. The following section outlines the most important of these conditions.

PREREQUISITES FOR NUCLEAR DISARMAMENT

The balance between the risks and benefits of comprehensive nuclear disarmament will be determined first and foremost by the overall evolution of the international political system. If deep animosities persist between major powers, if their governments are seen as unstable, unaccountable, or inclined toward treachery, or if technically capable states continue to challenge international norms of behavior, the balance will remain unfavorable. If, on the other hand, the major powers enjoy good relations, if their decision making processes and military deployments are reasonably transparent, if they have confidence that other states will abide by international norms, and if they are willing and able to take collective action to counter aggression, the prospects for prohibiting nuclear weapons will be greatly improved. The committee does not wish to imply that comprehensive nuclear disarmament would require the creation of a global utopia, but neither would it deny that a substantial positive evolution in international politics will be required. U.S. policy can play a significant role in helping this favorable evolution take place, but it must be borne in mind that the necessary changes will take time. The changes cannot be mandated, and in a nuclear disarmed world they must apply to all states, not just the present nuclear weapons states.

The elimination of armed conflict between states is not a precondition for the prohibition of nuclear weapons. Although Article VI of the NPT calls for a "treaty on general and complete disarmament" in connection with nuclear disarmament, this is neither a necessary nor a sufficient condition. States will not agree or adhere to a prohibition on nuclear weapons unless they are confident their vital interests could be adequately protected without such weapons. A fundamental attribute of sovereignty is the ability to defend oneself, whether this be through national resources alone or through alliance systems or other international means. The committee believes that serious efforts should be made to achieve comprehensive international arrangements to regulate conventional force structures and deployments at the lowest levels consistent with national and international security interests and at the lowest costs to the world economy. Such arrangements are beyond the scope of this study, but they would go a long way toward reducing the risk of conventional conflicts.

Comprehensive nuclear disarmament will require a highly effective system of verification to confirm that all nuclear weapons had been dismantled and that all fissile materials had been placed under international safeguards. The system would have to provide timely warning of any attempt to build new nuclear weapons or to reconstruct dismantled nuclear arsenals. Most or all of the required inspection procedures and surveillance capabilities would be developed in the course of reducing national nuclear arsenals to the level of a few hundred warheads, and in the course of improving the effectiveness of International Atomic Energy Agency (IAEA) inspections. The main difference is that states are likely to demand an increasing degree of confidence in the proper functioning of verification systems as the number of nuclear weapons is reduced to zero, which will require an unprecedented level of cooperation and transparency among all technically capable states.

In support of a regime prohibiting nuclear weapons, technical means of verification could be supplemented by national and international laws making it a crime for any individual knowingly to participate in the development, production, acquisition, transfer, or use of nuclear weapons, together with measures designed to increase the probability of "leaks" or "whistle blowing" by those who may be aware of such activities. A comprehensive nuclear disarmament treaty could, for example, require parties to enact laws obligating citizens to report any information about possible violation of the treaty to the international inspection agency and make it illegal for states to retaliate against whistle blowers. Such measures could be particularly valuable in uncovering activities that are difficult to detect, such as the concealment of nuclear weapons or weapons materials. At least some individuals involved in a covert illegal national program might be expected to report such activities.

As long as nuclear power and other peaceful nuclear activities continue, there will be a risk that associated materials and facilities could be diverted to military purposes. The proper management and structure of civilian nuclear activities therefore will be of central importance in a nuclear disarmed world. The first nuclear disarmament proposal, the Baruch Plan, proposed by the United States in 1946, envisioned the creation of an "International Atomic Development Authority" that would control all mining, refining, and distribution of uranium; own all facilities capable of producing fissile materials; and inspect and license all other nuclear activities.[5] Although an agency with this scope and authority would be impractical today, given that most nuclear facilities are privately owned and operated, some aspects of the nuclear fuel cycle that are especially worrisome could be limited or brought under international control. Stocks of weapons-usable fissile materials,[6] as well as facilities that produce or use such materials (particularly enrichment and reprocessing), could be managed by an international agency. In addition, fuel cycles could be modified to increase barriers to the diversion of these materials and to decrease or possibly eliminate the production and use of fissile materials in forms directly usable in nuclear weapons.

Although the committee did not examine verification issues in detail, two points seem obvious. First, no conceivable verification regime could, by technical means alone, obtain high confidence that it had accounted for every nuclear weapon or every kilogram of fissile material that had been produced. It could not be ruled out that a former nuclear weapons state had kept a few "bombs in the basement" or enough fissile material to build a few weapons. Second, the inherent capability of many states to build nuclear weapons would make it difficult to provide timely warning of an attempt to do so, particularly if fissile materials were diverted from civilian facilities.

It is possible that these considerations would prove to be relatively unimportant. For example, if relations among all major states were as cooperative as are today's relations among the United States, the United Kingdom, and France, one might not worry about the possibility of "bombs in the basement" or breakout among those states with maximum potential for rapid and massive revival of nuclear capabilities. Moreover, if the decision making processes of these governments were sufficiently transparent, states might judge that the probability that bombs or weapons-grade fissile materials could be hidden from inspectors for many years was negligible. It seems more likely that the potential for cheating or breakout would be regarded as cause for significant concern, in which case the disarmament regime would have to incorporate safeguards to deter and deal with these possibilities.

Safeguards might include security guarantees that pledge states to aid victims of nuclear attack or to punish nations that attempt to build, brandish, or use nuclear weapons; international nuclear or conventional forces of sufficient strength to deter, prevent, or punish the use of nuclear weapons; or preparations to rebuild national nuclear forces should the verification system detect violations. If collective security arrangements were strong, as measured by political will and military ability to punish violators, or if states believed that any advantage that could be obtained by violating the agreement would be short lived (e.g., because other states would quickly rebuild their arsenals), incentives to cheat or break out would be small.

As noted above, assessing the establishment of robust and comprehensive collective security arrangements or international military forces is beyond the scope of this report. The committee will, however, elaborate on one possible type of safeguard that has received considerable attention: maintaining the ability to rebuild national nuclear arsenals.

Any agreement prohibiting nuclear weapons would have to specify what constitutes a nuclear weapon, and which activities related to nuclear weapons would be permissible and which would not. A continuous spectrum of weapons-related activities is possible under a prohibition, ranging from theoretical and experimental work on nuclear problems, to the construction and operation of civilian nuclear facilities, to sustaining an ability to design and fabricate nuclear weapons, preserving facilities for this purpose, and, in the extreme case, retaining stockpiles of weapons components. There are advantages and disadvantages to setting the demarcation line near either end of this spectrum.

Several authors have argued that allowing countries to maintain a capability to build nuclear weapons in a short period of time would strengthen the nuclear deterrent effect, thereby permitting nuclear weapons to be prohibited without requiring major changes in the international order.[7] In this scenario, weapons-related facilities, activities, materials, or components would be placed under international monitoring. An attempt by any state to retrieve these components or use these facilities would trigger alarms in other nuclear-capable countries, leading them to assemble and disperse their nuclear weapons. The knowledge that any attempt to break out of the disarmament agreement would produce a rapid and offsetting response by other states would deter cheating in the first place, because cheating could produce no lasting advantage. Allowing states to maintain the capacity to rebuild nuclear weapons also would diminish the incentive for states to keep a few concealed nuclear weapons as a hedge against the possibility that other states might do the same. Under the regime of permitted activities, it might be necessary to protect the weapons-building capacity of each state against preemptive attack by other states, through a combination of multiple sites, deep burial, or provisions for rapid dispersal.

There are two potential problems with this type of arrangement, however. First, allowing states to maintain the capability to build nuclear weapons on short notice would make it easier for a state to cheat while at the same time making it more difficult to detect cheating. Permitted weapons-related activities would be of great value for a clandestine program and would create a background of legal activity against which it would be more difficult to detect illegal activities. Second, having states poised to resume manufacture and deployment of nuclear weapons could create dangerous instabilities in which states might rush to rearm during a crisis, thereby worsening the crisis. Drawing the demarcation line closer to the other end of the spectrum would simplify verification, allow more time to respond to signs of breakout, and build a larger firebreak to nuclear rearmament.

This discussion illustrates the importance of ensuring the stability of a comprehensive nuclear disarmament regime. If, in a crisis or other foreseeable circumstances, a prohibition on the possession of nuclear weapons created incentives to cheat or strong pressures to rearm, the risk of nuclear war could be higher under disarmament than with small national arsenals. In order for the balance of risks to favor moving to comprehensive nuclear disarmament, the three factors mentioned above—international politics, verification, and safeguards—must interact in ways that do not create such perverse incentives or pressures. Unfortunately, it is not possible to be much more specific without knowing more about the political and technical circumstances in which comprehensive nuclear disarmament would be pursued.

ROUTES TO NUCLEAR DISARMAMENT

The risks and benefits of comprehensive nuclear disarmament also would be affected by the way in which the transition away from small national arsenals is implemented. The committee has considered a number of possible means to achieve a prohibition on the possession of nuclear weapons and does not mean to suggest that these approaches are the best or only ways to deal with the challenge. Any such proposal would require extensive study by the states themselves and intensive negotiations among them over an extended period. When the time came, the nuclear weapons states and other states might find some other arrangement more appropriate to the conditions and norms of international politics then in existence. The committee's treatment of possible routes to prohibition is thus necessarily exploratory, in contrast to the analysis in previous chapters that resulted in specific recommendations.

Any option for achieving a durable prohibition on nuclear weapons must address a number of fundamental questions about how the transition from small national nuclear arsenals to total prohibition would be managed and how such a regime would operate:

1. *Who owns nuclear weapons during the transition?* Would they be controlled by the existing nuclear weapons states? Might the nuclear weapons states create a multilateral organization? Or would a truly international organization be responsible for these residual capabilities?

2. *Who controls nuclear weapons during the transition?* Would nuclear weapons remain under the operational control, however circumscribed, of the current nuclear weapons states? Would some joint or cooperative arrangements be developed to share responsibility? If control would pass to an international body, who would belong to that body and how would decisions be made?

3. *Would the authority to use nuclear weapons be part of the regime's mandate?* If so, under what circumstances might they be used? How would decisions to use nuclear weapons be made? How would the possibility of use be made credible?

4. *Who would maintain nuclear weapons capabilities?* Who would oversee the cadre of technically knowledgeable people charged with maintaining the safety and reliability of the remaining weapons?

5. *In what sequence would warheads and delivery vehicles be dismantled?* If survivability remained a critical factor during the transition, how would the balance be struck between that and the need for reassurance and verification?

Very broadly, the committee notes two major approaches to managing the transition to complete nuclear disarmament, each of which has a number of possible variants. One possible path for managing the transition to comprehensive nuclear disarmament would involve having an international agency assume joint or full custody of the arsenals remaining during the transition to prohibition. Alternatively, nations might find it preferable to bypass the intermediate step involving an international agency and proceed directly to negotiations to prohibit nuclear weapons either globally in a single agreement or in steps involving successive expansions in the number and geographical scope of nuclear weapon free zones.

In their current conceptual state, neither option can provide convincing responses to all of the questions posed above. Each tries to address a particular set of problems among the many that would have to be resolved if the world were to embark on an effort to prohibit nuclear weapons. Together they illustrate the strengths and weaknesses of different approaches, as well as the amount of effort and creativity that would be needed to make comprehensive nuclear disarmament a practical enterprise.

Option I: International Control of Nuclear Weapons

A transition to comprehensive nuclear disarmament could be managed by having an international agency assume full or joint custody of remaining nuclear stockpiles. A new agency could be created for this purpose, or the IAEA could be expanded and given this mission. This approach would be designed to ensure that the remaining nuclear weapons would no longer be instruments of national policy. During the transition, nuclear weapons under international custody would serve the core function of deterring the threat or use of nuclear weapons that might be retained or acquired by renegade states.

The membership of the agency, and the mechanisms by which it would reach decisions, would be the subject of much study and negotiation. The current nuclear weapons states undoubtedly would want a major role in the operation of the agency in return for agreeing to its creation. A key issue would be the degree of consensus that would be needed to take action. A balance would have to be achieved between ensuring that decisions regarding nuclear weapons enjoyed very broad support, and giving particular states or a small group of states veto privileges.

After the agreement establishing the agency entered into force, the agency would assume custody of all remaining nuclear weapons as well as all nuclear weapons-usable materials. Note that custody implies a legal responsibility but not necessarily physical possession, operational control, or ownership. In fact, custody of the weapons and materials could be managed in several ways.

One method for managing the warheads would be the design of a "dual-key" control system. Each nuclear weapon would be placed under the joint control of the international agency and the nuclear weapons state in physical possession of the weapon. The fire control system of the weapon would be modified so that the weapon could not be used or readied for use without the explicit approval of both the international agency and the owning state. Implementing such a system appears to be technically feasible.

Another method would have the nuclear weapons states move all their nuclear weapons into internationally safeguarded enclosures on their territory. The removal of warheads from these enclosures would require either concurrence of the agency or, at a minimum, would alert the agency that a withdrawal had taken place. This method is roughly analogous to the physical control maintained by the United States over nonstrategic weapons placed under the operational control of European NATO commanders.

A third method would be to transfer ownership and operational control of all remaining nuclear weapons to an international agency, which would be under the authority and command of the United Nations Security Council. This international nuclear force would be responsible for managing, maintaining, and, if necessary, using nuclear weapons. The use of nuclear weapons would be authorized only in response to the actual use of nuclear weapons by a state; there would be no authorization to threaten to use nuclear weapons to counter any other transgressions. This would preserve the core function of nuclear weapons—deterrence of the use of nuclear weapons—without the risks associated with continuing national control. At some point, the Security Council could determine that this deterrent function was no longer needed, at which time the international nuclear force could be disbanded and its weapons dismantled.

The establishment of such a force was envisioned by the Baruch Plan in 1946; it provided for temporary international custody of nuclear weapons pending their destruction once international controls were in place. With U.S.-Soviet relations deteriorating rapidly and the Soviets engaged in a major program to build a nuclear weapon as quickly as possible, the proposal did not lead to any agreement. Now that the Cold War is over, security concepts proposed at the end of World War II might finally find acceptance, albeit in a world where technical and military capabilities are far more widely diffused than they were 50 years ago.

It is difficult to define today the circumstances under which the nuclear weapons states would transfer to an international organization full authority over the control and use of nuclear weapons. Such an act would presuppose a degree of confidence in international organizations, and a level of trust and cooperation between major powers, that would seem to make the deterrence provided by the international force unnecessary. Although an international force would ameliorate the risks associated with national control of nuclear weapons, it would raise a whole new set of questions, not the least of which is "Who would police the policeman?"

Option II: Prohibit Nuclear Weapons by Direct Diplomatic Process

Nations might find it preferable to bypass the intermediate step of transferring custody of residual stockpiles to an international agency and proceed directly to a prohibition on nuclear weapons. One route would be to convene an international conference of the five nuclear weapons states with the goal of agreeing to eliminate their nuclear arsenals according to a specified schedule. A convention among the nuclear weapons states, supplemented by the provisions of the NPT, would establish a worldwide legal framework prohibiting nuclear weapons.

A convention limited to the five nuclear weapons states would be relatively straightforward and uncomplicated, but failure to include other states in the negotiation could compromise international support for the resulting agreement and would forego the opportunity to negotiate constraints on nuclear proliferation beyond those contained by the NPT. In addition, a five-power agreement would have to include some mechanism for addressing the problem of the undeclared states. As indicated in Chapter 3, it is possible that one or more of the undeclared states might eliminate their nuclear arsenals and join the NPT in advance of a universal prohibition on the possession of nuclear weapons, particularly if progress is made toward resolving the regional security concerns of these states. If undeclared states remain at this stage, the process of bringing them into the prohibition regime will have to be structured so that a declaration of their nuclear capabilities would not lead to instability or otherwise impede progress toward disarmament.

A second approach to comprehensive nuclear disarmament would be an international conference charged with creating a new treaty to prohibit the possession of nuclear weapons. This new treaty would replace the NPT and possibly other treaties such as the Comprehensive Test Ban Treaty. The negotiations leading to the Biological and Chemical Weapons Conventions are examples of an international process to outlaw an entire class of weapons. Although this approach would undoubtedly take much longer than a five-power negotiation, it would have the advantages of engaging all states as equal partners and of permitting an opportunity to create additional nonproliferation measures, such as anytime-anywhere inspections and restrictions on the production or use of weapons-usable materials, that go well beyond the NPT.

A third approach would be to convene a conference to amend the NPT to prohibit the possession of nuclear weapons by all parties. Under Article VIII of the NPT, amendments must be ratified by a majority of all NPT parties, which must include all of the nuclear weapons state parties and all parties that were, at the time, members of the IAEA Board of Governors. This approach would have the advantage of instantly capturing all of the states now party to the NPT, but the problem of the undeclared nuclear states would remain.

A fourth approach to prohibition would be through an expansion in the number and geographical scope of nuclear weapon free zones (NWFZs). When all of the current and new NWFZs are in force, nuclear weapons will be prohibited in all of the southern hemisphere (except the oceans) and in significant portions of the northern hemisphere. Frequently suggested candidates for additional NWFZs include Central Europe, the Middle East, South Asia, and Northeast Asia. By negotiating additional NWFZs that include regions of potential conflict between nuclear-armed states and, ultimately, all nuclear-armed states, a global prohibition on the possession or use of nuclear weapons could be achieved in a piecemeal fashion.

The committee cannot predict when, whether, or under what conditions the nuclear weapons states and undeclared states would be willing to accede to a regime that, under any of the proposals suggested above, would require the elimination of their nuclear arsenals. The assumption here is that these states, having agreed in earlier years to unprecedented transparency measures and reductions, would be more prepared than at present for such a step. The world of several decades hence is still malleable, and future initiatives by the United States and Russia could well make the transition to comprehensive nuclear disarmament much less visionary and uncertain than it looks from the present vantage point.

CONCLUSIONS

Achieving the conditions necessary to make a durable global prohibition on the possession of nuclear weapons both desirable and feasible will not be easy. Complete nuclear disarmament will require continued evolution of the international system toward collective action, transparency, and the rule of law; a comprehensive system of verification, which itself will require an unprecedented degree of cooperation and transparency; and safeguards to protect against the possibility of cheating or rapid breakout. As difficult as this may seem today, the process of reducing national nuclear arsenals to a few hundred warheads would lay much of the necessary groundwork. For example, the stringent verification requirements of an agreement on very low levels of nuclear weapons and fissile materials might by then have led to some new or expanded international agency with vigorous powers of inspection.

The potential benefits of comprehensive nuclear disarmament are so attractive relative to the attendant risks—and the opportunities presented by the end of the Cold War and a range of other international trends are so compelling—that the committee believes increased attention is now warranted to studying and fostering the conditions that would have to be met to make prohibition desirable and feasible.

NOTES

1. The time that would be required for a country to build (or rebuild) a nuclear arsenal depends on many factors, including the country's level of technical and industrial development, the existence of nuclear facilities and materials that might be available for a weapons program, the presence of scientists and engineers with expertise in nuclear weapons design and manufacture and/or the existence of programs to maintain the capability to build nuclear weapons, the desired number and sophistication of the weapons, and the degree of urgency and priority accorded to the effort and the level of resources devoted to it. In the Manhattan Project the United States accomplished the remarkable feat of building two different types of fission weapons in less than three years. If all nuclear warheads were eliminated, the current nuclear weapons states, and probably a dozen or more other countries, could in a national emergency produce a dozen simple fission bombs in as little as a few months, even if no effort had been made to maintain this capability. On the other hand, the production of a hundred lightweight thermonuclear bombs or warheads equipped with modern safety and security devices might take several years, even if special efforts had been made to maintain the capability to produce such weapons.

2. International Court of Justice, "International Court of Justice: Advisory Opinion on the Legality of the Threat or Use of Nuclear Weapons," *International Legal Materials*, vol. 35 (1996), p. 830.

3. Ibid., p. 831.

4. There is some ambiguity, of course, in what constitutes "war between major powers." Major or great powers are defined by their relative military, economic, and industrial strength, and by their interest, involvement, and influence in interstate politics and security; interstate wars generally are defined as conflicts resulting in a significant number of battle deaths (e.g., more than 1,000 per year). If one includes the Korean War, in which China fought against a coalition led by the United States but neither side declared war on the other, the succeeding period (which now stands at 45 years) is longer than any previous period of peace (or absence of war) between the major powers. See Jack S. Levy,

War in the Modern Great Power System: 1495-1975 (Lexington: University of Kentucky Press, 1983).

5. U.S. Department of State, *Documents on Disarmament, 1945–1956* (Washington, D.C.: U.S. Government Printing Office, 1960), pp. 10–15.

6. In this context, "weapons-usable fissile materials" are materials that could be used in a nuclear weapon without further enrichment or reprocessing. This includes separated plutonium of any isotopic composition and highly-enriched uranium, as well as unirradiated compounds or mixtures containing these materials.

7. See, for example, Jonathan Schell, *The Abolition* (New York: Avon, 1984); and Michael J. Mazarr, "Virtual Nuclear Arsenals," *Survival*, vol. 37, no. 3 (Autumn 1995), pp. 7-26.

APPENDIXES

APPENDIX
A

Biographical Sketches of Committee Members

JOHN P. HOLDREN (NAS member), chair, is Teresa and John Heinz Professor of Environmental Policy and director of the Program in Science, Technology, and Public Policy, John F. Kennedy School of Government, and professor of Environmental Science and Public Policy in the Department of Earth and Planetary Sciences at Harvard University. He is also a member of the President's Committee of Advisors on Science and Technology, Chair of the Executive Committee of the Pugwash Conferences on Science and World Affairs, Visiting Distinguished Scientist at the Woods Hole Research Center, and a consultant to the Lawrence Livermore National Laboratory. He has written extensively on energy technology and policy, global environmental problems, and international security.

JOHN D. STEINBRUNER, vice-chair of CISAC, is a senior fellow and former director of the Foreign Policy Studies Program at the Brookings Institution. He has held faculty positions at Yale, Harvard, and Massachusetts Institute of Technology. A political scientist, he has written extensively on arms control and security issues, including problems of command and control and crisis decision making.

MAJOR GENERAL WILLIAM F. BURNS (USA, ret.), chair of the nuclear weapons study, was the ninth director of the U.S. Arms Control and Disarmament Agency and former deputy assistant Secretary of State for Political-Military Affairs. He served as the first U.S. Special Envoy to the denuclearization negotiations with states of the former Soviet Union under the Nunn-Lugar Act, and he negotiated the government-to-government agreement on HEU sales to the United

States of uranium from dismantled Soviet weapons. He is currently a distinguished fellow at the U.S. Army War College.

GENERAL GEORGE LEE BUTLER (USAF, ret.) is former commander-in-chief of the Strategic Air Command and its successor, United States Strategic Command. Prior to assuming these positions, he served as director for Strategic Plans and Policy on the staff of the Chairman of the Joint Chiefs of Staff. He played a leading role in adapting U.S. national military strategy and nuclear war planning to the post-Cold War era. He is now president of Kiewit Energy Group, with headquarters in Omaha, Nebraska. In 1996, he served as a member of the Canberra Commission on the Elimination of Nuclear Weapons.

PAUL M. DOTY (NAS) is director emeritus of the Center for Science and International Affairs and professor emeritus of the Department of Biochemistry and Molecular Biology at Harvard University. He was a member of the President's Science Advisory Committee and has served as a consultant to various government agencies. He has been a leader in developing dialogues on security issues between Russian and American scientists.

STEVE FETTER is an associate professor in the School of Public Affairs, University of Maryland. A physicist, he was a special assistant to the Assistant Secretary of Defense for International Security Policy and a Council on Foreign Relations fellow at the U.S. Department of State.

ALEXANDER H. FLAX (NAE) is president emeritus of the Institute for Defense Analyses and served as home secretary of the National Academy of Engineering. His field is aeronautical engineering. From 1964 to 1969 he was assistant secretary for research and development of the U.S. Department of the Air Force and has served on advisory boards of the U.S. Departments of Defense and Transportation.

RICHARD L. GARWIN (NAS, NAE, IOM) is fellow emeritus of the Thomas J. Watson Research Center of the IBM Corporation. An experimental physicist, he has served as a consultant to the Los Alamos National Laboratory on nuclear weapons and to the U.S. government on topics of national security and arms control. He has been a member of the Defense Science Board and is the 1996 recipient of the R.V. Jones Intelligence Award.

ROSE GOTTEMOELLER is deputy director of the International Institute for Strategic Studies. Previously she was director of Russian, Ukrainian, and Eurasian affairs at the National Security Council and a senior specialist on Soviet security policy for the RAND Corporation.

SPURGEON M. KEENY, JR. is president of the Arms Control Association. He served with the atomic energy division of the U.S. Department of Defense, as a senior staff member of the National Security Council, and with Arms Control Disarmament Agency as assistant director for science and technology (1969-1973) and as deputy director (1977-1981). He was head of the U.S. delegation to the Theater Nuclear Force Talks in 1980.

JOSHUA LEDERBERG (NAS, IOM), chair of CISAC's Working Group on Biological Weapons Control, is university professor and past president of the Rockefeller University. In 1958 he received the Nobel Prize in Physiology or Medicine for his work in bacterial genetics. He has been active in the work of the National Science Foundation and National Institutes of Health and was involved in National Aeronautics and Space Administration Mariner and Viking missions to Mars. He served as a consultant to Arms Control Disarmament Agency during the negotiation of the Biological Weapons Convention.

MATTHEW MESELSON (NAS, IOM) is professor of molecular and cellular biology at Harvard University and codirector of the Harvard Sussex program on Chemical and Biological Warfare, Armament, and Arms Limitation. He has served as a consultant on chemical and biological weapons matters to U.S. government agencies.

WOLFGANG K. H. PANOFSKY (NAS), chair emeritus, is professor and director emeritus at the Linear Accelerator Center, Stanford University. He served as chair of CISAC plutonium study. His field is experimental high-energy physics. He was a member of the President's Science Advisory Committee under Presidents Eisenhower and Kennedy and the General Advisory Committee on Arms Control to the President under President Carter.

C. KUMAR N. PATEL (NAS, NAE) is vice chancellor of research at the University of California at Los Angeles. He is the former executive director of research, Material Science, Engineering, and Academic Affairs Division of AT&T Bell Laboratories. He has served as a trustee of Aerospace Corporation, Los Angeles and director of California Micro Devices, Milpitas, Calif. He is currently a director of Newport Corporation and chairman of the board of directors of Accuwave Corporation. He has received several awards in the field of lasers and quantum electronics. Most recently he was awarded the National Science Medal by President Clinton in 1996.

JONATHAN D. POLLACK is senior advisor for international policy at the RAND Corporation. A political scientist, he is a specialist on East Asian political and security affairs, especially China. He has served as a consultant to the Ford

Foundation and Los Alamos National Laboratory and is a member of the National Council on U.S.-China Relations.

REAR ADMIRAL ROBERT H. WERTHEIM (USN, ret.; NAE) was director of navy strategic systems projects from 1977 to 1980, responsible for development and support of U.S. submarine launched ballistic missile systems. From 1981 to 1988 he served as Lockheed Corporation's senior vice-president for science and engineering. He is a member of various advisory groups serving, among others, the Department of Defense, Department of Energy, U.S. Strategic Command, and the University of California.

APPENDIX
B

The Buildup and Builddown of Nuclear Forces

Chapter 1 of this report discusses the evolution of the world's nuclear forces during the Cold War and the development of the constraining influences on that evolution. This appendix presents data describing this rise and fall in graphical form.

Unfortunately, the information available to support graphical summaries of this kind from unclassified official U.S. government sources is only fragmentary. Reproduced here, therefore, are data on U.S., Soviet/Russian, British, Chinese, and French forces from the Natural Resources Defense Council (NRDC).[1]

There are considerable uncertainties in these figures, due to definitional ambiguities, disagreements among sources, and basic lack of information. Tabulation includes "on-line" forces, irrespective of their alert status, and those off-line, that is, in repair or modification. Nonoperational units and test units are not included. Naturally the data on Soviet/Russian, French, Chinese, and British forces are based on estimates, with sources frequently disagreeing. The references from which these graphs are taken contain an extensive discussion of sources and numerous qualifications about the reliability of the data.

Fortunately, precision in these numbers is not required to make a number of broad observations:

- During the buildup, the United States led the Soviet Union in numbers of nuclear weapons by 6 to 10 years.
- The peak buildup rate in the nuclear weapons stockpiles, particularly that of the United States (about 5,000 weapons per year) was substantially larger than the currently feasible dismantlement rate (about 1,500 to 2,000 weapons per year).

- U.S. total megatonnage declined steadily from the mid-1960s (and remained considerably below Soviet megatonnage). The average yield of the U.S. nuclear weapons declined from its peak above 1 megaton to just above 200 kilotons today.
- The number of nonstrategic (tactical) nuclear warheads has declined much more sharply than that of strategic warheads, but Russian tactical warheads are expected to remain more numerous than those of the United States.
- The total stockpiles of both Russia and the United States today remain above the 10,000-warhead level.
- While a tabulation of Soviet nuclear megatonnage is not included in this appendix, total Soviet megatonnage remained considerably higher than that of the United States in the latter part of the Cold War.

NOTE

1. The data on U.S. and Soviet/Russian forces are taken from Robert S. Norris and Thomas B. Cochran, "Nuclear Weapons Databook: U.S.-U.S.S.R./Russian Strategic Offensive Nuclear Forces, 1945-96" (Washington, D.C.: Natural Resources Defense Council, January 1997). The figure on British, Chinese, and French forces is created from NRDC data that appear as a regular feature in the *Bulletin of the Atomic Scientists*. These data are from "Nuclear Notebook: Estimated Nuclear Stockpiles 1945-1993," *Bulletin of the Atomic Scientists*, vol. 49, no. 10 (December 1993), p. 57.

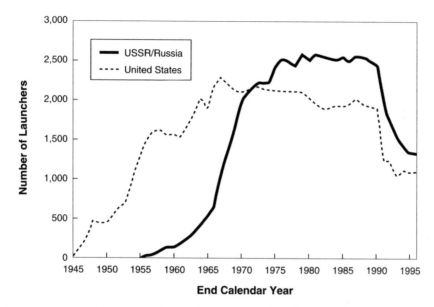

FIGURE B.1   U.S.-USSR/Russian total strategic launchers (force loadings), 1945-1996.
Source: Natural Resources Defense Council.

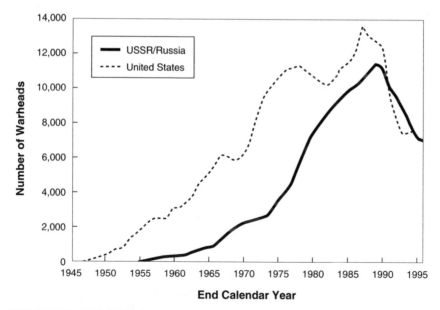

FIGURE B.2   U.S.-USSR/Russian total strategic warheads (force loadings), 1945-1996.
Source: Natural Resources Defense Council.

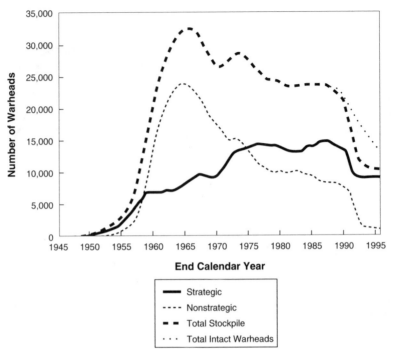

FIGURE B.3   U.S. nuclear stockpile, 1945-1996.   Source: Natural Resources Defense Council.

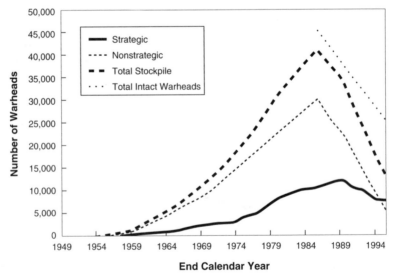

FIGURE B.4   U.S. -USSR/Russian nuclear stockpile, 1949-1996. Source: Natural Resources Defense Council.

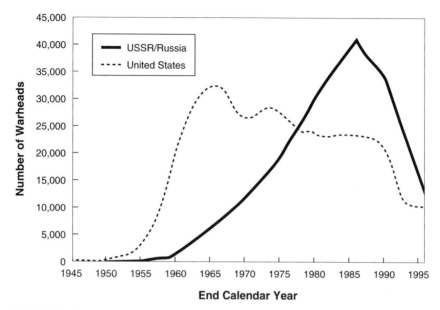

FIGURE B.5   U.S.-USSR/Russian nuclear stockpile, 1945-1996.   Source: Natural Resources Defense Council.

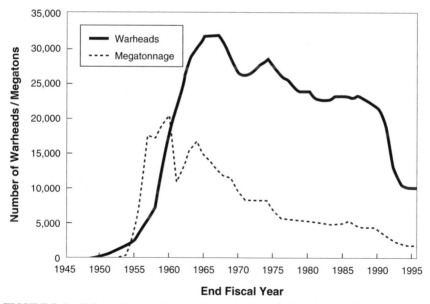

FIGURE B.6   U.S. nuclear warheads and megatonnage by fiscal year.   Source: Natural Resources Defense Council.

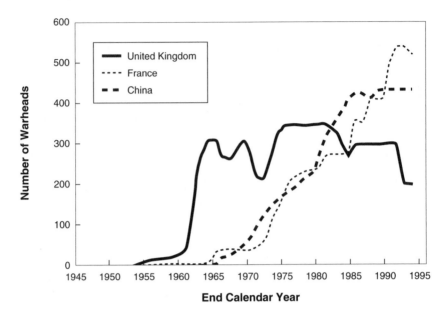

FIGURE B.7   Estimated nuclear weapons stockpiles of the UK, France, and China, 1950-1993. Source: Data provided by the Natural Resources Defense Council.